Interdisciplinary Approaches to the Study of Mysticism

Series Editors
Thomas Cattoi
Santa Clara University
Berkeley, CA, USA

Bin You
Minzu University
Beijing, China

The exploration and interpretation of mystical phenomena is an integral part of the study of religion and spiritual practice, which consistently attracts the interest of scholars and the general public. At the same time, the term "mysticism" may encompass all kinds of transformative practices leading to an experience of ultimate reality or the divine outside the context of particular religious traditions. As a result of the increasingly interdisciplinary character of the study of humanities, scholars are becoming more interested in the contributions of different academic disciplines to the understanding of mystical phenomena. In the spirit of this growing conversation across disciplinary boundaries, the series provides a space for the interdisciplinary study of mysticism, where new methodologies informed by psychology, the natural sciences, or the humanities complement more traditional approaches from religious studies and theology. The series also privileges interreligious and comparative approaches to the study of mysticism, with a particular interest in Asian religions and minority religious traditions.

More information about this series at
http://www.palgrave.com/gp/series/14851

June McDaniel

Lost Ecstasy

Its Decline and Transformation in Religion

palgrave
macmillan

June McDaniel
Department of Religious Studies
College of Charleston
Charleston, SC, USA

Interdisciplinary Approaches to the Study of Mysticism
ISBN 978-3-030-06512-6 ISBN 978-3-319-92771-8 (eBook)
https://doi.org/10.1007/978-3-319-92771-8

Cover image: Leonid Plotkin

Printed on acid-free paper

This Palgrave Macmillan imprint is published by the registered company Springer International Publishing AG part of Springer Nature
The registered company address is: Gewerbestrasse 11, 6330 Cham, Switzerland

This book is dedicated to Dr. Robert Garvin, my undergraduate philosophy professor and friend. I find that in this book I am dealing with the same problems of the nature of religious experience and its effects on culture that he discussed long ago, inspired by his own classes with Paul Tillich. His interests in Asian religions motivated me to enter the field of History of Religions.

Contents

CHAPTER 1

Introduction: What Happened to Ecstasy? Mysticism, Ecstasy, and the Constructivist Loop

The study of mystical and ecstatic experience is out of fashion in the modern field of Religious Studies. Analysis of religious consciousness has been obscured by interest in politics and sociology, in conflict and power structures. This was particularly visible at the 2017 meeting of the American Academy of Religion, at a session on Huston Smith and perennialism.

All cultures have both similarities and differences, but the study of similarities came in for some heavy condemnation. Studying common themes was condemned by the speakers are "boring," "dated," "pointless," in general a waste of time. It was something not worth doing any more, only suited to scholars from a century ago. What was modern and interesting was the study of differences, especially those that led to conflicts and wars. This was what was exciting. There were some speakers in the session that spoke in favor of the perennial approach, but they could not match the visceral hatred of religious experience shown by the speakers who condemned it. This was shown in the sneering laughter at the "wreck of the good ship Eliade," and statements of relief that we are no longer living in the "bad old days" of looking for commonalities in religious experiences that occur to people in different cultures. Indeed, sympathy and interest in the experiences of other cultures was described by several of the participants as "so old-fashioned."

Academic fields often reflect the cultures of their writers, and modern Western culture clearly emphasizes differences as well. Political parties, social groups, and religions themselves are at high levels of conflict and

J. McDaniel, *Lost Ecstasy*, Interdisciplinary Approaches to the Study of Mysticism, https://doi.org/10.1007/978-3-319-92771-8_1

fragmentation, a process which has been under critique since the 1960s.[1] The breakdown of society into various sorts of competing factions comes to be the major focus of much of modern Religious Studies.

In this book, I choose to describe something unfashionable. I would like to bring the field back to the study of religious consciousness, at least partly by documenting how much of the field opposes it. The modern study of ecstatic religious consciousness over the last thirty to forty years has largely been a study of objections to its subject matter. I believe there is a need for counter voices, to respond to this change.

We see this in Theology as well as Religious Studies. The general Religious Studies response to mystical experience has been that it is impossible to study in itself (the neo-Kantian view), that is largely irrelevant to politics and society (the Marxist view) or that it should be left to the theologians. But the theologians don't want it either—as scholars, they are historians, linguists, ethicists and textual interpreters, as well as voices for social change. The study of mystical and ascetical Theology, along with metaphysics generally, has been largely de-emphasized in modern seminaries. Like the religionists, theologians have largely shifted their interests to the social and political world, often substituting classes in practical skills like small-business organization, finance, leadership, and preaching skills for areas like ascetical and mystical Theology. Ecstasy is the "hot potato" that neither religion nor Theology wants.

As the Benedictine monk David Steindl-Rast has recently noted, "Every religion seems to begin in mysticism and end in politics."[2] He compares mystical states to the hot lava of a volcano, and organized religion to the dry crust and ash that forms as it cools, when it settles and loses energy. In a similar way, he notes that the volcanic passions of mystical states turn into the organized religious institutions that show the symptoms of "rigor mortis." This is not merely to say that human experiences inevitably come in contact with concrete material and social life, but that they become the dominant aspects of religious understanding.

In this approach, he follows William James. In 1902, James wrote in his *Varieties of Religious Experience* about the contrast between religion as a "dull habit," and its role as an "acute fever."[3] The "dull habit" role has become quite popular among scholars of both religion and mysticism, with the emphasis on culture, language, and politics which limit intense religious experiences to prescribed methods of expression and predetermined contexts. As we have noted, the study of James' "acute fever" or "hot place" has been largely rejected. From the interest in religious origins, found in early writers like Rudolf Otto and Friedrich Scheleiermacher, we have moved to the political ends, found in critical theory and the current interest in the influence of politics and economics on religious activities. While James saw the acceptable religious states supported by religious institutions as secondary and imitative, we now see the reverse, where the mystical state itself is viewed as derivative, an imitation of current political and social models. Study of the mystical state as the origin of religions thus becomes a historical artifact, a quaint reminder of nineteenth-century scholarship.

During the early twentieth century, scholars of religion considered the exploration of ecstatic religious experience important for the study of religion. But over the last thirty years or so that approach came in for sharp criticism. Many scholars wanted to get away from issues like ecstasy and mysticism because they seemed to suggest an experiential essence of religion, which they wished to avoid, and turned instead to analyzing discourses about experience, power, and privilege. This problem is well-documented in Leigh Eric Schmidt's article "The Making of Modern Mysticism" in his 2003 article in the *Journal of the American Academy of Religion*. He calls mysticism a "beleaguered category," subject to a fall from theoretical grace due to the recent emphasis on the body and the material world in the field of Religious Studies. Writers in the field who are more interested in a materialistic or naturalistic view of the world have no compunction about insulting the category—mysticism is an "essentialist illusion," a "false category," a strategy to protect religious claims from scientific investigation and a way of avoiding political and ethical concerns. It is "sunk in disrepute," nothing but a fabricated category which is only an evasion of more important categories, such as analyses of power relationships. We see mystical and ecstatic experiences described as regressive, narcissistic, morally inferior, colonizing, solipsistic, and delusional.[4]

This is the sort of hostility that used to be found in theologians talking about heresies. However, the modern writers attacking mystical and ecstatic states work from a different set of assumptions. It is not that essentialist or perennialist claims come from beliefs about the wrong religion, but rather that there can be no truth to any possible religious claims about ecstatic and mystical states, for there is nothing to be found beyond history, culture, language, and the body. Schmidt looks at some eighteenth-century critiques of ecstasy, when it was associated with enthusiasm, the experience of the "god within." He notes many examples of horror at mystical and ecstatic states, perhaps the most memorable by Bishop William Warburton, who compared such states to the "excremental waste" of reasonable religion.[5]

This horror has found new modern incarnations in critical theory and the hermeneutic of suspicion. However, while these approaches are popular in academic circles, social behavior does not always follow academic values. According to a recent Pew research poll, almost 50% of Americans reported that they had a mystical or ecstatic experience. In 2009, the Pew Forum on Religion and Public Life released a report on spirituality in America. Entitled "Many Americans Mix Multiple Faiths," the report points out that many Americans are now choosing to "blend Christianity with Eastern or New Age beliefs" and that "sizable minorities of all major U.S. religious groups" have said that they have had supernatural experiences. For the first time in 47 years of polling, the number of Americans who said that they have had a religious or mystical experience was greater than those who said that they had not had such an experience. It notes:

> Approximately half of Americans (49%) say they have had "a religious or mystical experience - that is, a moment of religious or spiritual awakening." This is roughly the same as the number that said this in 2006 (47%), but it represents a sharp increase over the past four decades. In 1962, only 22% of Americans reported having had such an experience, which grew to about a third in 1976 (31%) and 1994 (33%). Since then, the number has continued to increase to roughly half of the public in this decade.[6]

Mystical and religious experiences are particularly common among the "religiously unaffiliated," among whom 51% have reported having a religious or mystical experience. This is quite a striking number of people who do not have a dominant religious commitment or model through which to interpret such experiences. This category "sunk in disrepute" needs another look; any academic theory which claims that there is a pathology that afflicts half of the individuals in a society should be subjected to considerable scrutiny.[7] Religious Studies scholars cannot justify ignoring the experiences of half the population when they claim mysticism is an false category and therefore unworthy of study. Mysticism and ecstatic religious experiences are not outliers to be easily ignored, devalued, and explained away.

We may also note that in the shorter Pew Research Survey published in 2016, "Americans may be getting less religious, but feelings of spirituality are on the rise," religious affiliation has not grown, but spiritual experiences have. The data compared a 2007 research survey with a 2014 one, on "America's Changing Religious Landscape." The Pew Survey specifically asked respondents about whether they have had a feeling of deep spiritual peace and well-being once a week (a number that rose from 52% in 2007 to 59% in 2014), and whether they feel a deep sense of wonder at the universe (which rose from 38% in 2007 to 45% of respondents in 2014).[8] Religious affiliation was evaluated by standard measures such as how important respondents say religion is to them and their frequency of religious service attendance and prayer. The authors state that "The growth of the unaffiliated population and their decreasing religiosity have been the main factors behind the emergence of a less religious public overall. But, interestingly, the rise in spirituality has been happening among both highly religious people and the religiously unaffiliated." As the authors note (with some surprise), the rise in deep wonder at the universe among atheists rose 17 points, from 37 to 54% of the respondents.

There is also a Daily Spiritual Experience Scale, developed in 2002 by Lynn Underwood and Jeanne Teresi, which studies the frequency of spiritual experiences. They examine the issue from a more psychological perspective, noting that such experiences are linked to decreased total alcohol intake, improved quality of life, and positive psychosocial status.[9] However, their focus is on the frequency rather than the intensity of such experiences, and their interest is in the psychological and social impact of such states. The fact that spiritual experiences have beneficial social effects is another reason why it is important to study them.

Some modern understandings of spirituality and ecstatic experience have been explored by Paul Heelas and Linda Woodhead in their book *The Spiritual Revolution: Why Religion is Giving Way to Spirituality.* They describe spirituality as a major cultural shift, a turn away from life lived in terms of external or objective roles, duties, and obligations, and "a turn towards life lived by reference to one's own subjective experiences."[10] It describes the older understanding of life as part of tradition and community, following higher authorities and accepting the necessity of conforming to a religious ideal. In this more conservative approach, the individual will and perspective must be sacrificed or subject to authority. This contrasts with the modern emphasis on emotions, feelings, individual conscience, and qualities like compassion, sensitivity, having the courage to be one's own authority, finding an individual path in life, and finding a connection with one's deeper self.[11]

Heelas and Woodhead cite a Gallup Poll from 2000, in which eight out of ten Americans express a desire for spiritual growth.[12] A third of the respondents "defined spirituality with no reference to God or a higher authority." To the question, "Do you think of spirituality more in a personal or individual sense, or more in terms of organized religion and church doctrine?" we find that 72% chose "personal or individual sense."[13] Indeed, even the definition of God is debated, as a spirit or life force, as a personal being, as a "higher power," as a deeper Self, as liberation, or the person's highest ideal. These are large numbers for Religious Studies researchers to ignore.

In his book *Mystical Sociology*, Philip Wexler calls spirituality "a contemporary form of mysticism that is less institutionally embedded."[14] He critiques modern sociology for being "despiritualized and demysticized" and notes how the study of spirituality tends to focus on its commercialized forms as New Age consumerism. Thus, we see modern mystical groups analyzed as institutions and commodities rather than for the inner experience they promote. They are castigated as "elite," and as "ideological weapons in class warfare," using methodologies such as literary and textual theory, psychoanalysis, French feminism, and political theory, which have become popular among sociologists. Mysticism becomes reinterpreted as solipsism, delusional omnipotence, egotism and self-obliteration which diminishes practical reason.[15] He notes that this is part of the broad neglect of the "vertical dimension" of human experience by the field. Post-secular society needs not only a resacralization, but a "remysticization," returning the concept of the sacred to both organized discourse and institutional life.[16] He focuses his critique on the field of sociology, but this would apply to the field of Religious Studies as well.

He notes that the denial of religious experiences in other cultures works as a part of the process of delegitimation of such experiences in our own culture. He finds that this delegitimation and lack of mystical presence in modern Western culture results in isolation, objectification, and reification. He finds these are shown in quests for euphoria (through drugs and alcohol), flights back to nature (through the popularity of health food and camping), secondary intimacy (encounter-type therapies), thrill-seeking (fast driving, shock movies), and hedonism (buying on credit, playboy-style sexuality).[17] He also notes that it is odd how mystical consciousness becomes less important in religions as it becomes more important in other fields, such as neurology and psychiatry.

This denial has been noted by some conservative theologians as well, who are disturbed at the prevalence of ecstatic states, in which "everything is possible and nothing is unbelievable." Craig Kennet Miller compares mystical experiences and supernaturalism with "the black market in a totalitarian country" for they make up "the vast underbelly of the American religious economy."[18] Religious ecstasy is thus widespread but illegitimate, and unadmitted by most religious authorities. He notes that this search for the mystical divine spark "is ignored by mainline Protestants, railed at by conservative fundamentalists, and strangely enough, embraced in some ways by Pentecostals and charismatics."

My own field of study has been the History of Religions, with a specialization in Hinduism, especially the bhakti, yoga, and tantric traditions of West Bengal. I have spent years interviewing living ecstatics and translating Bengali biographies and autobiographies of holy people. It is quite striking how ecstasy is highly valued there, while in the West in recent years it has been condemned and largely ignored. We still see visionary experience, mystical union, and possession trance in Bengal, and these are major criteria for who becomes accepted as a *siddha*, a perfected person, who may become a guru if he or she chooses to teach.

Indian universities in general do not have departments of Religious Studies, though we do see a few programs in specific religions focusing on history and literature, and we do currently have a few universities trying some experimental comparative religion programs. India is a democracy with freedom of religion, and part of freedom of religion has been keeping it out of the schools. The lack of religion programs has had both advantages and disadvantages.

A major disadvantage is that it is difficult to evaluate gurus, religious leaders, and teachers. While some are sincere, others are mercenary and concerned with dominance over their followers. It is difficult to argue against them when there are no critical standards that students are taught which can be used to judge them. Another problem is that there are large religious minorities in India (various forms of Muslim, Sikh, Jain, Parsi, and now Tibetan Buddhist groups) and Hindu students are not taught their basic beliefs and practices. These minority communities often live separately from people following Hinduism, and this separation is increased by religious groups having their own separate legal systems for personal law. On issues like marriage, divorce, and inheritance, there are different religious standards for different communities. When people are viewed as totally different, in practice and law and often appearance, it is easy to view them as alien, and easy for riots, "communal violence" and wars to start. In this situation, the critical distance taught in the field of Religious Studies can contribute to religious and ethnic groups better understanding each other.

On the other hand, India has not gone through the politicization in the field of Religious Studies that we see in Western scholarship. In the nineteenth century, there were groups like the World Parliament of Religions exploring interfaith dialogue, and scholars who emphasized areas of shared belief and practice as a way to create harmony and understanding between religious communities. In its early years, the field also had researchers with an interest in religious consciousness and experience, and the ways that these were understood in different cultures. Both goals were intended to contribute to interfaith dialogue, and they have continued to support respect for religious experience among many Hindu groups.

The goal of reaching world peace by finding harmony between religions was a goal for many years. However, in the later twentieth century, Religious Studies took a reverse direction. It became opposed to finding similarities, with the only legitimate techniques of study used to focus on differences.[19] I saw this directly at the time of Mircea Eliade's death, when I had just come to the University of Chicago to study History of Religions. I was told directly that there was only one Eliade, and he was dead—from now on the field would only focus on area studies. There would be no more broad comparative studies, and I should forget this "consciousness nonsense" and study languages.

The only acceptable universal or comparative model remaining is Marxism via critical theory, with the only commonality between persons and communities being power and the desire for economic and political dominance. The idea that there are no cultural universals except for power, which has a special and unassailable status, is difficult to justify rationally and is often defended with almost religious intensity. The focus on conflict at various levels has meant that the original areas of interest, such as religious emotion, revelation, and the vertical or depth dimension of religion, have become unacceptable and illegitimate areas of study. This has been related to arguments over the category of religion itself, and whether it is still relevant for the modern study of religious events. These arguments over definition largely avoid questions of experience.

Scholars who value the study of power and conflict in religion often suppress other methods of inquiry, and with this suppression, the field can descend into yellow journalism, studying the sleazy underbelly of religion rather than its values and ideals. There is, of course, no lack of sleazy underbelly to focus on, from the history of colonialism to the oppression of heretics to transnational capitalism. The question is whether these are the only legitimate areas of study in the field of religion.

While the study of religion in India lacks the comparative categories of understanding and analysis that we see in Western departments, it has also been free of the Western reduction of religion to political and economic forces. In India, these ideas are present, but mostly in its Communist groups, such as the CPI, the CPIM, and the Naxalites (Communist Party of India, Marxist-Leninist). They have often worked with the Indian Rationalist group, which has sought to portray traditional religion as superstition. I first ran across the Rationalists in West Bengal, where renunciant sadhus that I interviewed told me of being chased from villages by children throwing rocks. They said that the children were taught by the local Rationalist group to keep sadhus away, telling children that sadhus were ignorant and wanted to keep them poor. For the Bengali Communist groups, religion was a set of beliefs that wasted time and money, ignoring the issues of power and privilege that were really important.

Here in the West, we see similar critiques. The modern field of Religious Studies has moved from a more sympathetic focus on experience to a more suspicious one. The lurching from one extreme to the other leaves many students unable to study what they wish to explore, and this is one reason for the shrinking of Religious Studies class sizes in Western colleges, and the growth of New Age and alternative spirituality. It is also a reason for the rise of the "spiritual but not religious" category that has been showing up on national surveys and polls.[20] This group has seekers with a great interest in religious experience, but not many places to learn about it. As this book will note, neither academic nor theological schools want to focus on this. Thus, we have the rise of alternative schooling, from televised self-help speakers on spirituality and happiness to shamans initiated at weekend workshops. These are groups on the other side of the spectrum, which talk about ecstasy and use religious language, but often have little academic background on the topic. These attempt to compensate for the lack of study of religious experience in the academy.

One would think the field of mysticism within Religious Studies would keep the interest in mystical and ecstatic consciousness alive, but it has been slow going. I have been on the Steering Committee for the Mysticism Group at the American Academy of Religion since 2000, off and on, and spent several years as co-chair of the group. During that time, we have sent out many calls for proposals on ecstatic and mystical religious experiences. We have received many fine proposals (which later became excellent papers) on the historical, biographical, literary, aesthetic, psychoanalytic, feminist, social, and political aspects of such experiences. However, the calls for trance, possession, altered states of consciousness, and other more experiential topics generally did not receive enough proposals to create sessions. The Mysticism Group has moved from a primarily Christian focus (which it had when I first joined it) to a more cross-cultural and comparative religion perspective, which was a very good way to expand the field. However, it has had a hard time moving toward the study of the current "acute fever" side, which has largely been left to physicians and neurologists.

There have been many reasons for the avoidance of the more ecstatic side, in favor of the focus on the social and cultural impact of religion. One part of the problem is that the field of Religious Studies has largely given up the study of religious consciousness and experience. The issue of consciousness has been left to psychologists, who have also been avoiding it. As a colleague in our college Psychology department phrased it, "When I go to conferences in my field, and someone mentions consciousness, people look around as if a bad smell had entered the room." It is not a suitable topic for scientific research, as science is limited to material events. The study of ecstatic consciousness is not considered to be "scientific." But is this the case?

The astronaut and aeronautical engineer Edgar Mitchell, famed for his work on the Apollo 14 mission, had an experience of unity which he called a "life-changing epiphany" during his return from a mission to the moon. He described it as accompanied by feelings of "ecstasy and bliss." When he sought to understand what had occurred, he ran into the same problem:

> A most fundamental difference between science and the cultural cosmologies is that science avoided the issue of "what is consciousness" during the four hundred years after philosopher Rene Descartes quieted the Inquisition by declaring that body and mind belonged to different realms of reality, thus allowing science to arise as a materialist pursuit in the Western world apart from religious influence and persecution. As a result, most science has assumed during those four centuries a reductionist approach and considered consciousness to be an epiphenomenon of molecular complexity, with evolution itself considered a random process of nature...[21]

Mitchell ended up trying to unite religion and science through his Institute for Noetic Studies, and trying to find in quantum theory a new approach linking science and ecstatic experiences. It was his view that "at the base of all evolved thought systems are concepts of transcendence," which religions have controlled through meditative and other practices.[22] He felt that this was an important area for science to explore.

The understanding of the goals of science is another problematic area. While science is a method of learning for many people, for others it is an end in itself, and a justification for rejecting other forms of knowledge. As the physicist Erwin Schroedinger, a Nobel laureate, noted in interview:

> The scientific picture of the real world around me is very deficient. It gives a lot of factual information... [but] it cannot tell us a word about red and blue, bitter and sweet, physical pain and physical delight, it knows nothing of beautiful and ugly, good or bad, God and eternity. So, in brief, we do not belong to this material world that science constructs for us... the scientific world view contains of itself... not a word about our own ultimate scope or destination.[23]

Scientists are not all as limited as non-scientists assume them to be. In Nancy Frankenberry's *The Faith of Scientists in their Own Words*, a wide variety of scientists discuss their views (many are not materialists, though she includes Richard Dawkins and Stephen Jay Gould for variety). Albert Einstein's views have become widely known: "I am of the opinion that all the finer speculations in the realm of science spring from a deep religious feeling... I also believe this kind of religiousness... is the only creative religious activity of our time."[24] Science could bring him a sort of ecstasy:

> The most beautiful thing we can experience is the mysterious. It is the source of all true art and science. He to whom this emotion is a stranger, who can no longer pause to wonder and stand rapt in awe, is as good as dead, his eyes are closed. This insight into the mystery of life, coupled though it be with fear, has also give rise to religion. To know that what is impenetrable to us really exists, manifesting itself as the highest wisdom and the most radiant beauty which our dull faculties can comprehend only in their most primitive forms- this knowledge, this feeling, is at the center of true religiousness. [25]

Such ecstatic religious feeling was his motivation for research:

> ... I maintain that the cosmic religious feeling is the strongest and noblest motive for scientific research. Only those who realize the immense efforts and, above all, the devotion without which pioneer work in theoretical science cannot be achieved are able to grasp the strength of the emotion out of which alone such work, remote as it is from the immediate realities of life, can issue... it is cosmic religious feeling that gives a man such strength. A contemporary has said, not unjustly, that in this materialistic age of ours the serious scientific workers are the only profoundly religious people.[26]

The legitimacy of the study of consciousness for science is a major problem. Another is the debate over interpretations of religious experience. Ecstasy is generally associated with mysticism and religious experience, and both are contested categories. Examples of these debates over the meaning of these terms can be found in William Parsons' recent edited volume *Teaching Mysticism*.[27] Writers who described and analyzed ecstatic and mystical states from a positive perspective, like Carl Jung and Mircea Eliade, have come under widespread attack. These attacks come both from *ad hominem* arguments and from the fear that these writers might believe "something religious about religion," no matter how unspecified. Indeed, they are suspected of "cryptotheology," implying that somebody somewhere is secretly believing in something non-material, even if nobody is quite sure what it is.[28] They have become academic outlaws.

In some ways, analytic philosophers talking about ecstasy bear a resemblance to theologians talking about heretical beliefs. Ecstatic claims are called false or malicious, due to ignorance or delusion. Even Rufus Jones, the well-known Quaker writer on mysticism, dismissed the visionary and ecstatic aspects of religious practice, and only accepted divine union or contemplative intuition as legitimate. As he noted in 1909, there are no "pure experiences," for experiences are all produced within "social and intellectual environments."[29] This argument is still going on a century later, notably between Robert Forman on the side of perennialism and pure experiences (as "consciousness without an object"), and Steven Katz and Wayne Proudfoot on the side of constructivism, in which all ecstatic experiences are determined or shaped by social learning. Over the past thirty years, we have not had a great deal of new theorizing on ecstatic experiences in the field of mysticism, partly because the perennialist/constructivist debate has largely stalled research. The constructivists have been winning.

In brief, the constructivist approach assumes that all experience is structured by preexisting categories, judgments and thoughts, that we have no access to any "uninterpreted given" or prereflective experience. As Wayne Proudfoot phrases it, "The experience is shaped by a complex pattern of concepts, commitments and expectations which the mystic brings to it… These beliefs and attitudes are formative of, rather than consequent upon, the experience. They define in advance what experiences are possible."[30] Steven Katz also emphasizes how experience is conditioned and interpreted in advance, with assumptions leading directly to experiences. He believes that "one cannot attain nirvana by accident."[31]

The problem in the argument is what I call here "the constructivist loop." Mystical and ecstatic states are usually defined and accepted by religions which have very specific doctrinal requirements. The experience must fit into doctrine—if it violates the beliefs, it is illegitimate, and thus called heresy, madness, or simply irrelevant eccentricity. Here is the problem of the constructivist loop—the experiences that are accepted by the tradition must always fit into the preexisting categories, thus all preexisting mystical categories can be said to define the legitimacy of the states. It is only a short step from there to arguing that they create the states. This seemingly unbreakable link between the doctrine and the mystical experience allows many religionists to claim that the doctrine is the cause of the mystical state. But in introducing mystical experiences which do not fit into doctrinal categories, this link is broken, and the causal relationship is shown to be false. When a limited range of religious beliefs determine what counts as data, this greatly narrows the range of possible evidence, and influences or determines the writer's conclusions.

Leaving out or denying data skews the full understanding of events, and false conclusions can result from such "self-selecting" data. I would compare the situation of omitting non-institutional ecstasy with a situation that I encountered as an undergraduate. At that time, the Psychology 101 courses required three hours as an experimental subject for all students in the classes. To get my three credit hours, I went to a large laboratory area where graduate students were writing down the names of students who would participate in the experiments. We were told that the experiments would involve giving and receiving electric shocks. I was horrified, and I told the bored graduate student writing down names that I would not participate in such an experiment. He irritably crossed off my name, and said, "Next." He did not ask for any reason for the withdrawal.

I later found out that these experiments were following the Milgram experiments at Yale University on obedience to authority figures, except that at in this case they were actually giving shocks (according to my boyfriend at the time, who received several electric shocks for his three Psychology credits). The researchers got comparable results to the Yale experiments—the majority of students undergoing the experiments were willing to shock others, even if they thought the shocks were harmful. However, the experimenters did not count those who refused to participate in the experiment when they first heard what was involved. I had refused, and there were other cross-outs on that list. But the papers by the faculty researchers analyzed only those participants who had already agreed to be in the experiment (we may note that several years of data and publications were thrown out years later on review by the university's Institutional Review Board for mistreatment of subjects).[32]

My point here is that the data were already limited before the researchers had started to analyze it. The sample was supposed to be broad and random, but it was really preselected. It is no surprise that, when the independent thinkers have been eliminated, those who follow authority remain. This is the problem with the constructivist loop; the pre-chosen data fit the model, and if it doesn't fit the model, it is excluded from study. Spontaneous and non-traditional ecstatic states are thus rejected as legitimate subject matter for analysis, as they do not reflect acceptable religious doctrine and literature. Such states have also been rejected by the field of Theology, following St. Paul's admonition to the Corinthians: "All things must be done decently and in order"

(1 Cor. 14.40). Non-institutionally sanctioned ecstasy is often disorderly and possibly indecent, and is thus not suitable for study in Theology.

So what should we do with the ecstatic states that contain religious content, which describe an experience of an Ultimate Reality or Truth or Absolute, but are not accepted by current religious institutions? The issue here is not that such experiences are similar or identical, but rather that they do not fit into cultural and religious preconceptions. Ignoring them has been a venerable tradition, as is viewing such states as hallucinations or some other form of pathology. It is an improvement over persecution for heresy, and having ecstatics institutionalized for schizophrenia or dissociative identity disorder. As Mary Douglas famously noted in *Purity and Danger*, things which do not fit into religious categories become defined as impure, polluted, even as abominations, as in the camel, hare, rock badger, and swine in the Levitical laws.

Religions accept experiences of the gods that they recognize. Ecstasy without specific institutional affiliation is largely ignored by the field of Religious Studies, but is considered not sufficiently sectarian by Theology. The Catholic Church has a long history of evaluating saints and their experiences to make sure that they are sufficiently doctrinal to be accepted. Heroic virtue will not count toward sainthood if the person's views are deemed heretical. There is a literature of attack from theorists of both perspectives. Ecstatic states are thus rejected by the fields in which one might expect interest.

Another reason for this neglect is the popular claim that mysticism and religious experience are modern, Western concepts, not found in older, non-Western cultures.[33] As they are not broad or universal categories of understanding and are only colonial impositions, they are thus not very important. Scholars would only be studying their own projected colonial categories when studying such states. We shall go into the problems with this argument in further detail in Chapter 8.

Do people need preexisting religious categories in order to experience transcendent ecstatic states? I found during fieldwork in West Bengal that the ecstatics and holy people that I interviewed tend to wander, seeking religious institutions into which to fit their experiences.[34] While some holy people became gurus, eventually adopting recognizable gods and rituals and gaining social acceptance in this way, others remained permanent wanderers, and never affiliated with any social or religious organizations. Yet others built ashrams which became centers of new movements, with their own rituals and deities. Many became charismatic leaders of new groups, with their following based on their experiences rather than any existing institutional organizations or doctrines.

Experiences of Bengali holy people are popularly judged by their impact on the observer—can a devotee, or even a stranger, share in the feelings of the mystic? In rural areas, ecstasy often determines legitimacy. And if the person is deemed legitimate based on the power of his or her ecstasy (*mahabhava*), the state of *divyonmada* or divine madness may allow the person to violate traditional religious doctrine and ritual. As one sadhu phrased it, *bhava* (religious or emotional experience) is stronger than *sastra* (religious literature and its beliefs). Indeed, some sadhus interviewed referred to themselves as *asastriya*, deliberately violating or ignoring religious literature and tradition. But if the person's inner state cannot be emotionally shared by others, and a religious explanation cannot be constructed to fit his or her experiences, then we have the popular interpretation of the state as *paglami*, ordinary or secular madness. With this interpretation, the person cannot gain disciples, the major criterion for becoming a religious teacher and leader, and people feel sorry for his or her family.

For some of the informants that I interviewed, the deity and guru lineage are a part of the ecstatic experience itself. For instance, I interviewed Gaudiya Vaishnava practitioners who had visions of the God Krishna, the saint/avatar Caitanya Mahaprabhu, his companion Nityananda and other Vaishnava saints. Such experiences would fit the constructivist model. Though their visions may not exactly have fit into Vaishnava doctrine, they were close enough to be recognizable. However, other informants had visions of light or the sense of mysterious power or energy or universal consciousness, and it was not clear which religious group or deity would be the most appropriate fit for their experiences. Some went from one ashram to another, trying to figure out if there was a group which would be suitable. The majority of those that I interviewed ended up as *sadhus* and *sadhikas* (male and female renunciants) within the universalist wing of Bengali Shaktism.

Such ambiguous experiences are not unique to India. Back in graduate school, a fellow student told me of his experiences of universal light and his efforts to find a religious interpretation that could accept or explain his experience. He had first identified it with Christianity and was pursuing a seminary degree in order to find an accurate understanding. When I met him, he was finishing the degree, but had already determined that his experiences did not fit into Christian categories (and the seminary was much more interested in ethics and preaching skills than the sort of experiences it deemed overly individualistic, and without suitably Christian content).[35] He had recently converted to Judaism, but it seemed that his experiences did not really fit there either, and he was thinking about following Islam.

He did not fit the stereotypes of the "New Age" dilettante—he was serious as a scholar and wanted to be part of a religious community. But there seemed to be no religion which included experiences like his. Like many modern Western spiritual seekers, he understood the spiritual experience and the social institution to be separate, and he gave the experience priority. For the constructivist approach to religious experience, what happened with him would be impossible. Yet he was quite sincere and faced a serious challenge to find religious community.

The field of Religious Studies today tends to focus on the political, literary, historical, and psychoanalytic aspects of religion. But when we look at issues like ecstasy, the anthropological and phenomenological sides are often neglected. I have interviewed ecstatics and visionaries in many countries, and they tend to be largely apolitical and ahistorical (as an example, many of the Bengali sadhus with whom I spoke were unfamiliar with the current president of India, World War II, the partition of India and Pakistan, and which other countries bordered India).[36] It was their intense experiences which guided their lives, ignoring the values of the social theories so popular with academic observers.

But today we see questions as to whether such intense experiences, within or without religious affiliation, are legitimate areas of study. Studies of non-traditional mysticism and spirituality can be seen in Wuthnow's "spirituality of seeking," Forman's "grassroots spirituality," Parson's "unchurched mysticism," Fuller's "unchurched America," Heelas' "inner-life spirituality" and "New Age Romanticism," Partridge's "re-enchantment," and the sort of widespread evangelical sense of presence described in T.R. Luhrmann's "When God Talks Back." We can see the more skeptical, and hostile, view from Proudfoot, Katz, McCutcheon, Fitzgerald, Penner, and other writers who interpret mysticism through the constructivist lens. It is an ongoing debate.

It is important to study mystical and ecstatic states for many reasons. As Ann Taves notes in her article "Religious Experience and the Divisible Self," studying unusual experiences can give us insight into the nature of religion as a category and unite a variety of fields as William James' research did:

> Openness to contested phenomena, such as the unusual experiences studied by James, and openness to comparing experiences understood as religious with experiences understood as nonreligious or even pathological, as advocated by Myers, allow us to explore the construction (and deconstruction) of religious experiences and, thus, the construction and deconstruction of religion as a category. Such work is inherently interdisciplinary and is one way… to help "reestablish the connection that has been lacking between religious studies and the other human sciences."[37]

Of course, we may note that given the modern necessity of pigeonholing any ecstatic experience into existing institutional or religious categories, classic comparative works such as James' *Varieties of Religious Experience* would likely have had much more difficulty being published if written today.

Ecstatic and mystical experiences can also give insight into creativity and give a broader notion of the self in relation to a much larger universe than ordinary consciousness and religious restrictions will allow. However, Schmidt's evaluation of mysticism as a fabricated and "beleaguered category" is still accurate for much of the modern academic study of religion.

One reason for such "beleaguerment" is the difficulty in agreeing on a definition for our terms. While there is a broad range of definitions for both ecstasy and mysticism, for this book we shall have religious ecstasy as William James' "acute fever," the intense and individual experience of the transcendent or Absolute, with no necessary institutional context. Thus, *ecstasy is an intense, altered state of emotion, perception, or identity which can be understood in relation to an ultimate goal or object.* It might be called the experiential aspect of mysticism. There are ecstatic states of union with the cosmos which are not part of any particular religious system, and there are ecstatic states which develop out of the practices of a religious system. Ecstatic states may involve a clearly delineated deity or religious goal, but do not require one. Nor do they require an institutional context. But if they are to be called religious ecstasy, they need to find something close enough to call a community. The ecstasies which become accepted by religious institutions then get improved titles—prophecy, revelation, or the word of God. The ecstasies which are not accepted tend to be forgotten, labeled as schizophrenia, Devil possession, or merely eccentricity. But there is ecstasy outside of religious institutions, which is how a book on atheist mysticism could make sense (the existence of ecstatic practices for atheists may be seen in a recent book title, *The Spiritual Atheist's Guide to Mystical Experiences and How to Have Them*).[38]

The term "mysticism" is a complex one—we have mystical Theology, mystical doctrine, mystical people, mystical texts, and mystical hermeneutics. The term "mysticism" can refer to a process or a goal, an exploration or a set of doctrines. But for our definition of mysticism here, we shall focus on the presence of a religious context. Thus, we define mystical ecstasy with a focus on interpretation, as *an intense, altered state of emotion, perception, or identity that is part of a complex of religious symbolism, and is recognizable to and valued within a religious tradition.*

Narrowing down the definition for ecstasy is important, as in this culture it has a wide range of associations and understandings. To show the real range of associations about a topic, one cannot surpass a thesaurus as a source. Here is the collection of synonyms for ecstasy from *Moby Thesaurus II*:

> abandon, affection, affectionateness, afflatus, amativeness, amorousness, ardency, ardor, beatification, beatitude, bewitchment, blessedness, bliss, blissfulness, cheer, cheerfulness, cloud nine, craze, delectation, delight, delirium, demonstrativeness, dharana, dhyana, ecstasis, ecstatics, elation, enchantment, enjoyment, euphoria, exaltation, excitement, exhilaration, exuberance, felicity, fervency, fervidness, fervor, fire, fire and fury, frenzy, furor, furore, fury, gaiety, gladness, glee, goatishness, gratification, gusto, happiness, heart, heartiness, heat, heaven, high spirits, horniness, hypnosis, hysteria, impassionedness, inspiration, intoxication, joy joyance, joyfulness, liveliness, lovelornness, lovesickness, madness, orgasm, orgy, overhappiness, overjoyfulness, paradise, paroxysm, passion, passionateness, pleasure, rage, rapture, ravishment, relish, rhapsody, romanticism, samadhi, savor, sentimentality, seventh heaven, sexiness, soul, spirit, sunshine, susceptibility, tearing passion, thrill, towering rage, trance, transport, unalloyed happiness, vehemence, verve, warmth, warmth of feeling, yoga trance, zeal.[39]

The fact that a culture could use the same word for bliss, rapture, goatishness, madness, orgasm, paradise, hypnosis, samadhi, spirit, and towering rage is in itself a worthy area of study. But in this book, we are narrowing our focus.

It is the intent of this book to urge the field of Religious Studies to include non-institutional forms of ecstatic and mystical experience as areas of study. There are two reasons why this would be useful. One is that the culture is not becoming totally secularized, and science has not supplanted religion, as Freud had suggested would happen in *The Future of an Illusion*. Instead, we have intense experiences that result in new forms of spirituality, and the charismatic and renewal adaptations of existing religions.

We also see that the traditional limitations of the fields of Religion and Theology make it difficult for these fields to deal with newer forms of ecstatic experience. Recent research with the drugs MDMA, LSD, and psilocybin has returned to the original use of the psychedelics—as a means of healing psychological trauma. Modern research with these drugs has shown that they can often induce many sorts of ecstatic states that could easily fit into James' *Varieties of Religious Experience*.[40] They also show extraordinary promise in dealing with PTSD and other psychological problems.[41] But there has been little interest in Religion or Theology in either the ecstatic or the therapeutic effects.

This book will show both the uses and problems of ecstatic states in both the study and the practice of religion generally, but it will focus on religion in the modern West. It argues that when there is no room for ecstatic experience within religions, we will find that it shows up elsewhere—in violence, in sexuality, in drugs, in music, and other areas.

The first chapter gives the introduction and argument for the book. The second chapter gives a background range of examples of religious ecstasy, so that we can have some data in the beginning. Chapters 3 and 4 will go over a range of attacks from the academic world and the theological "circle of faith." The fifth chapter explores the role of destructive or transgressive ecstatic states, in which violent and horrific experiences substitute for religious ones. Here, we may see the negative impact of a wide selection of transgressive and destructive behaviors that may result from the denial of ecstasy, and why it is so important for the study of religion to understand such experiences. The sixth chapter observes backlash, from the so-called "spiritual" ecstasies, the transformation of once-sinful activities into spiritual paths. The seventh chapter, on the return of the repressed, looks at some modern charismatic and renewal responses to this rejection of ecstatic states by mainstream religions. The eighth chapter examines an example of academic denial of ecstatic experiences, specifically from Indian religion. The ninth chapter describes some older writers and new directions. The concluding chapter discusses the ways that the rejection of its own data is harming the field of Religious Studies, and some ways to get beyond the current approaches of trivializing, criminalizing, and pathologizing ecstatic states. Without some understanding of intense experience, both the study of religion and the cultural sphere on which individuals draw to think about their own lives will be diminished.

This book will note how the suppression of ecstasy (from theological invisibility to political and economic reductionism to the drug wars) guarantees a set of countercultures and other reactions which will seek it in various ways. It will conclude with some suggestions for bringing together the exiled ecstasy and its home in the study of religion.

This book is not a comprehensive study, which would be far too long for one volume. However, it can give insight into the high cost of rejecting or ignoring the study of ecstasy, and the reasons why the field of Religious Studies has become alienated from such study. Today, it may be accurately characterized as a field which is "experience-distant."

To avoid such distance, we begin with some examples, to show the breadth of religious ecstasy.

Notes

1. As opposed to the critiques of conformity of the 1950s, more recent analyses of Western society have been more concerned with alienation and hostility. Gary Marx gives a typical list of issues showing modern conflict and fragmentation: the rise of strident single issue political groups, a negative politics that earmarks candidates to be defeated rather than elected, a retreat from public institutions and the public provision of health and welfare services (e.g., increased attendance at private schools), the dispersion of political power from the national to the state and local level, high rates of predatory crime, inventions such as the personal radio and video games that can isolate the person from direct interaction with others, the appearance of medieval-like walled or "gated" towns (with access to the town itself gained only by passing elaborate security and surveillance measures), etc. See Gary Marx, "Fragmentation and Cohesion in American Society," in *Disasters, Collective Behavior, and Social Organization*, ed. R. Dynes and K. Tierney (University of Delaware Press, 1994).
2. David Steindl-Rast, "The Mystical Core of Organized Religion," *Revision* 12, no. 1 (Summer 1989): 11–14.
3. "It would profit us little to study this second-hand religious life. We must make search rather for the original experiences which were the pattern-setters to all this mass of suggested feeling and imitated conduct.

These experiences we can only find in individuals for whom religion exists not as a dull habit, but as an acute fever rather." See William James, *The Varieties of Religious Experience* (New York: Collier Books, 1974 (1902)), p. 24.
4. See Eric Leigh Schmidt, "The Making of Modern Mysticism," *Journal of the American Academy of Religion* 71, no. 2 (June 2003).
5. Bishop William Warburton, "The Doctrine of Grace: Or, the Office and Operations of the Holy Spirit Vindicated from Insults of Infidelity, and the Abuses of Fanaticism," 1763, cited in Schmidt, ibid., p. 279.
6. From the online source Pew Forum.org, at http://pewforum.org/Other-Beliefs-and-Practices/Many-Americans-Mix-Multiple-Faiths.aspx.
7. We shall discuss the pathologizing of ecstasy in the fields of Religious Studies and Theology in Chapters 3 and 4.
8. "Americans may be getting less religious, but feelings of spirituality are on the rise," by David Masci and Michael Lipka, Pew Research Center, January 21, 2016, at http://www.pewresearch.org/fact-tank/2016/01/21/americans-spirituality/.
9. Lynn G. Underwood, Ph.D. and Jeanne A. Teresi, Ed.D., Ph.D, "The Daily Spiritual Experience Scale: Development, Theoretical Description, Reliability, Exploratory Factor Analysis, and Preliminary Construct Validity Using Health-Related Data," *Annals of Behavioral Medicine* 24, no. 1 (2002): 22–33, online at http://www.dsescale.org/underwoodteresi.pdf, February, 2018.
10. Paul Heelas and Linda Woodhead, *The Spiritual Revolution: Why Religion Is Giving Way to Spirituality* (Malden: Blackwell, 2005), p. 2.
11. Ibid., p. 3.
12. Ibid., p. 49.
13. Ibid., p. 74.
14. Philip Wexler, *Mystical Sociology: Toward Cosmic Social Theory* (New York: Peter Lang, 2013), p. 17.
15. Ibid., p. 33.
16. Ibid., pp. 25–26.
17. Ibid., p. 32.
18. Craig Kennet Miller, *Baby Boomer Spirituality: Ten Essential Values of a Generation* (Nashville: Discipleship Resources, 1993), p. 73.
19. See the debates in K. C. Patton and B. C. Ray, eds., *A Magic Still Dwells: Comparative Religion in the Postmodern Age* (Berkeley: University of California Press, 2000). Further discussion of the legitimacy of comparative religion and its problems can be found in David M. Freidenreich, Comparisons Compared: A Methodological Survey of Comparisons of Religion, from 'A Magic Dwells' to *A Magic Still Dwells: Method and Theory in the Study of Religion* 16 (2004): 80–101.

20. According to a 2012 study conducted by Pew Research Center, the number of Americans who do not identify with any religion has increased from 15% in 2007 to 20% in 2012, and this number continues to grow. One-fifth of the US public and a third of adults under the age of thirty report themselves to be unaffiliated with any religion, but they identify as being spiritual. See Funk, Cary; Smith, Greg. "'Nones' on the Rise: One-in-Five Adults Have No Religious Affiliation" (PDF), pewforum.org. The Pew Forum on Religion and Public Life, 2012.
21. Edgar Mitchell, "Epiphany in Space and on Earth: Reflections on the Akashic Experience," in *The Akashic Experience: Science and the Cosmic Memory Field*, ed. Ervin Laszlo (Rochester: Inner Traditions, 2009).
22. Ibid., p. 228.
23. E. Schroedinger, "Why Not Talk Physics," in *Quantum Questions: Mystical Writings of the World's Great Physicists*, ed. K. Wilber (Boston: Shambhala, 2001), pp. 83–85. Cited in Marjorie Hines Woollacott, *Infinite Awareness: The Awakening of a Scientific Mind* (Lanham: Rowman and Littlefield, 2015).
24. Nancy H. Frankenberry, ed., *The Faith of Scientists in Their Own Words* (Princeton: Princeton University Press, 2008), p. 151. This quote is from Einstein in His Own Words, 1930–1940.
25. Ibid., p. 154. From Einstein's "What I Believe," *Forum and Century (1930–1940)* 84, no. 4 (October 1930): 192–193.
26. Ibid., p. 157. From the Einstein "Religion and Science" section.
27. William Parsons, *Teaching Mysticism (AAR Teaching Religious Studies)* (Oxford University Press, 2011).
28. It is reminiscent of H. L. Mencken's famous definition of Puritanism: "Puritanism: The Haunting Fear That Someone, Somewhere, May be Happy."
29. Rufus Jones, *Studies in Mystical Religion*, 1909, cited in, ibid., p. 291.
30. Wayne Proudfoot, *Religious Experience* (Berkeley: University of California Press, 1985), p. 121.
31. Ibid., p. 123.
32. This problem was not publicized at the time, but it was later mentioned in "SUNY Albany Admits Violations, Health Department Settles Case." *Statesman* [Stony Brook, NY]. October 28, 1977: 2. "The first case to arise under New York's 1974 law governing research with human subjects has been settled, with university researchers admitting 35 violations involving 'at least 975 subjects.'" Experiments were not properly authorized or supervised, and some subjects were endangered by faulty equipment such as an electric shock machine, officials of Albany State University admitted yesterday. [...] "The case at Albany came to light because of complaints by a psychology graduate student." Online at

http://minervawept.blogspot.com/2014/05/ualbanys-mistreatment-of-human-subjects.html.

33. This argument was made by Proudfoot, in his *Religious Experience*, and also Robert Sharf in his article "Experience." See R. H. Sharf, "Experience," in *Critical Terms in Religious Studies*, ed. M. C. Taylor (University of Chicago Press: Chicago, IL, 1998).
34. Data for this can be found in my book *The Madness of The Saints: Ecstatic Religion in Bengal*, ibid., Chapters 5 and 6.
35. Individuality was not emphasized at the seminary. Another MDiv student there was a soldier who believed that he had been called by God to become a minister. I spoke to him as he was first coming into the seminary, and he was very enthusiastic. But I met him again on the graduation line, where he told me that he had been wrong—God would never speak to a human being. The focus of Christianity was the group, not the individual. Indeed, he said that if his experience showed him something different from what the Bible taught, then his own experience was wrong, and he would not listen to his own senses and ideas. He had learned that the group has greater truth than the individual.
36. This lack of interest is quite different from Bengalis more generally, who are willing to speak about politics and history at length. Indeed, these are major topics of discussion at coffeehouses and tea stalls. But sadhus do not consider themselves to be living in ordinary time, thus worldly problems are secondary.
37. Ann Taves, "Religious Experience and the Divisible Self: William James (and Frederic Myers) as Theorists of Religion," *Journal of the American Academy of Religion* 71, no. 2 (2003), p. 320.
38. Soren Sorensen, *The Spiritual Atheist's Guide to Mystical Experiences and How to Have Them* (Spiral Garden, 2018).
39. *Moby Thesaurus II by Grady Ward, 1.0.* http://onlinedictionary.datasegment.com/word/ecstasy.
40. As examples, see K. A. MacLean, M. W. Johnson, R. R. Griffiths, "Mystical Experiences Occasioned by the Hallucinogen Psilocybin Lead to Increases in the Personality Domain of Openness," *Journal of Psychopharmacology* 25, no. 11 (November 2011): 1453–1461. https://doi.org/10.1177/0269881111420188. Epub 2011 Sept 28. Also, A. Garcia-Romeu, R. R. Griffiths, and M. W. Johnson, "Psilocybin-occasioned Mystical Experiences in the Treatment of Tobacco Addiction," *Current Drug Abuse Review* 7, no. 3 (2014): 157–164.
41. Michael C. Mithoefer et al. "The Safety and Efficacy of ± 3,4-Methylenedioxymethamphetamine-assisted Psychotherapy in Subjects with Chronic, Treatment-Resistant Posttraumatic Stress Disorder: The First Randomized Controlled Pilot Study," *Journal of Psychopharmacology* 25, no. 4 (April 2011): 439–452.

CHAPTER 2

Some Examples of Religious Ecstasy

We cannot have a book on the topic of religious ecstasy without giving at least a few examples. Here, we shall discuss some forms of ecstasy from five major world religions, with a sampling of the range of ecstatic states that have been accepted and valued by religious practitioners, and the ways that these states have been interpreted.

We may note that religious ecstasy is not necessarily an individual experience. From Durkheim's "collective effervescence" to Victor Turner's "communitas," ecstatic states can bind people together as they experience their gods and find deep joy and states of union. In West Bengal, one of the criteria for being accepted as a *siddha*, a perfected person or saint, is the ability to induce ecstatic experiences in others. Some saints could create "waves of *bhava*" which were said to drown whole villages in ecstasy, with hundreds of people laughing in joy, crying, dancing, having visions, and falling into trances. Religious ecstasy was described as a "sky-high typhoon" in which crowds danced wildly in states of possession.[1] We also see group possession among many of the Adivasi or tribal groups in India. Among the Oraons, for example, the Old Lady of the Grove jumps from one dancer to another during the spring festival, until all are possessed together, uniting them as members of the group.[2]

However, the most detailed accounts of such states tend to come from sources which describe individual experience, thus religious ecstasy is often understood to be an isolated event, occurring in solitude, as William James emphasized. Theologies which interpret religious ecstasy also tend to focus on the individual rather than the group. But we

J. McDaniel, *Lost Ecstasy*, Interdisciplinary Approaches to the Study of Mysticism, https://doi.org/10.1007/978-3-319-92771-8_2

should remember that ecstasy can occur in both individual and group settings.

Judaism

Judaism is not usually considered to be an ecstatic religion. In its modern mainstream forms, it is a religious of scripture and analysis, of obedience to divine revelation, of ethics and community. Language and philosophy are important in understanding God's words in the Torah; personal emotion and experience are less highly valued as ways to understand God's will. However, we do see ecstasy based on the Hebrew Bible: in the experiences of the prophets, and the stories that are understood by mystics to describe the creation of the universe and the ascent to God's heavenly throne.

In the Hebrew Bible, a major section is the Neviim or the books of the prophets. Prophets were believed to speak with God, and God would direct and guide them, though sometimes he would use them to threaten punishment to their people for having hard hearts, stiff necks, and general pride and disobedience. God was also believed to occasionally speak with judges, elders, and priests. This ability to directly interact with God is generally believed to have been lost in the past, usually about the time of the early days of the Second Temple.[3] The prophets wrote primarily during the ninth through sixth centuries BCE, giving speeches, diatribes, exhortations, poems, narratives, warnings, and threats. As a first-person account of prophetic ecstasy, we can look at Isaiah's vision in Isaiah 6.1:

> In the year of King Uzziah's death I saw the Lord seated on a throne, high and exalted, and the skirt of his robe filled the temple. About him were attendant seraphim, and each had six wings; one pair covered his face and one pair his feet, and one pair was spread in flight. They were calling ceaselessly to one another, "Holy, holy, holy is the Lord of Hosts: the whole earth is full of his glory"...[4]

Another example is Ezekiel's prophetic call in Ezekiel 1-3, sometimes called the "flying saucer vision" as he describes wheels within wheels descending out of the sky. In Kabbalistic tradition, this came to be known as the chariot or *merkabah*, with its four symbolic levels of wheels, creatures, firmament, and the Throne of God:

> On the fifth day of the fourth month in the thirtieth year, while I was among the exiles by the river Kebar, the heavens were opened and I saw a vision of God… I saw a storm wind coming from the north, a vast cloud with flashes of fire and brilliant light about it; and within was a radiance like brass, glowing in the heart of the flames… The wheels sparkled like topaz, and they were all alike in form and working they were like a wheel inside a wheel, and when they moved in any of the four directions they never swerved in their course… and high above all, upon the throne, a form in human likeness. I saw what might have been brass glowing like fire in a furnace from the waist upwards, and from the waist downwards I saw what looked like fire with encircling radiance… When I saw this I threw myself on my face, and heard a voice speaking to me: Man, he said, stand up, and let me talk with you.[5]

Ezekiel's vision is long and complex and has been subject to many later interpretations, but it is a good example of the kind of detailed perceptions that can make up a prophetic call.[6]

Scholars have debated whether such experiences can be described as mystical ecstasy, as some definitions require the person to merge his identity with God. Von Rad notes that prophets are often in a "condition of abnormal excitation," with the normal consciousness raised to an intensity never experienced in ordinary life, and they were led out to "new vistas of belief."[7] He states that the prophet is "drawn into the emotions of the deity himself … it was not only the knowledge of God's designs in history that was communicated to him, but also the feelings in God's heart, wrath, love, sorrow, revulsion… Something of Yahweh's own emotion passed over into the prophet's *psyche* and filled it to the bursting point."[8]

Prophetic vision usually involved imagery: angels, the heavenly temple, the descent of the throne of God to earth on storm-clouds, scrolls, almond rods, and seething cauldrons, as well as glimpses of future disasters. God is seen and heard, through altered perception and revelatory vision. The person becomes God's mouthpiece, his oracle, and messenger, but he does not unite his essence with God.

Abraham Heschel debates whether the prophets can be understood as ecstatics, as there is no Hebrew word for ecstasy in the bible. However, the Greek translation of the Bible or Septuagint uses the word "ekstasis" as a translation of "*tardemah*," a Hebrew word for sleep sent by Yahweh (as in Adam's sleep in Gen 2:21, when Eve was created from his rib). Philo tended to use Greek ideas to understand the prophets—he called them hierophants, who could enter states of divine possession (*entheos*) and frenzy (*mania*). In describing Moses he states a basic idea: "No prophecy without ecstasy."[9]

However, the rabbinic tradition had a suspicion of ecstatic states, and some writers distinguished the Biblical prophets from other mediums and diviners of the time by their lack of ecstasy—the prophets passed along ideas clearly and rationally, while foreign mediums would "chirp and mutter," get possessed, faint, go into frenzies, and enter other states which were social liabilities. However, not all prophets were clear and rational. Many prophets roamed in ecstatic bands, with altered states being contagious—one could enter a trance state by merely being in their presence. They might use loud music, dance, mutilate themselves, and lay naked in public.[10] An example of such problematic bands is seen in First Samuel. In this story, King Saul encountered Samuel and a group of prophets and caught their state of trance. As 1 Sam 19:20-24 states of Saul and his soldiers:

> When they saw the company of prophets in rapture, with Samuel standing at their head, the spirit of God came upon them and they fell into prophetic rapture… on his way there the spirit of God came upon him too and he went on, in a rapture as he went, till he came to Naioth in Ramah. There he too stripped off his clothes and like the rest fell into a rapture before Samuel, and lay down naked all that day and all that night. That is why men say, 'Is Saul also among the prophets?'

In this case, when people ask if Saul is also a prophet, it is often understood to question whether or not he is sane (the phrase is sometimes used colloquially today in the same way).

There are also writers on Biblical prophecy who have emphasized the importance of ecstasy. As Gunkel wrote, "The fundamental experience of all types of prophecy is ecstasy."[11] Weber also described the presence of ecstasy,

> Psychologically viewed, most pre-Exile prophets were ecstatic men. At least, Hosea, Isaiah, Jeremiah and Ezekiel professed to be and undoubtedly were. Without gross carelessness, one may safely assume that all were ecstatics, though of various kinds and different degrees.[12]

We also see ecstasy after the prophets, in the Jewish mystical traditions of Kabbalah. Here, we shall look at Abraham Abulafia, the thirteenth-century mystic and ecstatic who developed prophetic ecstasy through the combinations of Hebrew letters, chanting the variations of the Holy Name, and the interpretive system of *gematria*.

Kabbalah means tradition, specifically the tradition of mystical understanding in Judaism. There is a long history of Kabbalah in Jewish Mysticism, which Gershom Scholem dates from the first century BCE.[13] One of the regions where Kabbalistic thought developed most richly was Spain, especially in the Golden Age of Kabbalism in Spain in the thirteenth and fourteenth centuries. Scholem contrasts two styles of Kabbalistic thought of that time, the "ecstatic school" and the "theosophical school." He calls Abraham Abulafia "the outstanding representative of ecstatic Kabbalism" and his ideas "the culmination of the ecstatic school."[14]

Abulafia emphasized personal experience of ecstatic states and wrote instruction manuals for meditation. Though he had visions that could be characterized as prophetic and apocalyptic, he did not publicize them throughout the Jewish community. Instead, he focused upon a small group of disciples and taught them his set of mystical techniques, the Path of the Names, or prophetic Kabbalism.

According to Abulafia (1240–1291), the soul is bound to the body with spiritual knots, and meditation is the way to free it. These knots separate the individual from his or her divine nature, and they blind the soul to the world of God. The soul needs to see beyond natural forms, without being blinded by divine light. It needs objects of meditation which are able to open up the divine realm without overwhelming the soul. These are the letters of the Hebrew alphabet, which reflect the Name of God. Abulafia also called his meditative technique *Hokhmath ha-Tseruf*, the science of the combination of letters.[15] It is when these letters begin to flow into the mind that the person enters an ecstatic state. Abulafia calls this world of revealed letters "the true world of bliss."[16]

The contemplation of the Divine Name and its variants in the Hebrew letters opens the seals, and the ecstatic immerses himself in the pure forms of the letters, which impress themselves upon his soul. This is God's language which reshapes him and prepares him for the entrance of divine light, a radiance which "illuminates his thought and heals his heart."[17] As Scholem describes it:

> This is the stage of prophetic vision, in which the ineffable mysteries of the divine name and the whole glory of its realm reveal themselves to the illuminate. Of them, the prophet speaks in words which extoll the greatness of God and bear the reflection of this image. Ecstasy, which Abulafia regards as the highest reward of mystical contemplation, is not, therefore, to be confused with semi-conscious raving and complete self-annihilation... when the bolts are shot back and the seals taken off, the mind is already prepared for the 'light of the intellect' which pours in.[18]

The prophetic faculty is the highest stage of individual human mind, and in that state it encounters the universal Active Intellect, which it may perceive in the form of a teacher or other respected person. The ecstatic identifies with the guide or teacher, and thus indirectly with God.[19] Abulafia's interest in ecstatic Kabbalah began with his own sort of prophetic call:

> When I was thirty-one years old, in the city of Barcelona, God woke me from my sleep and I studied *Sefer Yezirah* with its commentaries; and the hand of God [rested] upon me, and I wrote some books of wisdom and wondrous books of prophecies, and my spirit was quickened within me, and the spirit of God came into my mouth, and a spirit of holiness moved about me, and I saw many awesome sights and wonders by means of these wonders and signs.[20]

In Abulafia's prophetic states, there are both fear and joy:

> ... all your body will begin to tremble, and your limbs will begin to shake, and you will fear a tremendous fear, and the fear of God shall cover you... And the body will tremble, like the rider who races the horse, who is glad and joyful, while the horse trembles beneath him.[21]

Abulafia's prophetic Kabbalism emphasizes the power of word and sound, which was based on combinations of Hebrew letters, numerical analysis as a guide to combining them,[22] and a study of the names of God. While it shows all three ecstatic qualities of intense emotion, altered perception, and changed identity, its focus was on perception and knowledge.

Ecstatics in the tradition of Kabbalah are understood to keep in touch with God, transmit messages, ask God for help for the Jewish people, and teach others through wisdom and selfless love. So long as there are holy people, the world will continue to exist.

Christianity

Ecstasy is present in many forms of Christianity, but here we will look briefly at Catholicism and Eastern Orthodoxy. Catholicism has a broad literature on mystical and ecstatic states written by church fathers and monastics. Saints, hermits, and mystics have gone into great detail on how such states occur, what varieties and types exist, and the sort of ritual practices that encourage them. As the Christian theologian August Poulain notes, ecstasy occurs

> when the divine action has a considerable force and all outside commun-ions with the senses are interrupted, or almost entirely so. Thus we are no longer capable of any movements such as are voluntary at least, nor are we able to come out of our prayer at will.[23]

Many Catholic writers have written on ecstatic experience, and one of the earliest and most influential has been fourth-century St. Augustine of Hippo. Augustine's definition of ecstasy is when "the soul's intention is completely turned away, or snatched away, from the body's senses."[24] The soul then does not perceive by means of the senses, it only sees by means of the imagination or intellect.

This distinction is based on Augustine's three categories of perception: corporeal (or sensory), imaginative (using mental imagery), and intellectual (a form of perception that does not use specific images or sensations). He developed these as a way of understanding St. Paul's visions, and as a response to the question of whether God can be seen with bodily eyes. He contrasted John's visions in the Apocalypse, which are imaginative, with the intellectual vision of pure light seen by Moses, and he notes that in this type of vision:

> There the brightness of the Lord is seen, not through a symbolic or corporeal vision… nor through a spiritual vision… but through a direct vision and not through a dark image, as far as the human mind elevated by the grace of God can receive it.[25]

The *ecstasis* of the intellectual vision is described here as in some sense direct and immediate, though Augustine has said that it is impossible to enjoy the vision of God in this life in a perfect and stable manner. Earthly visions will have less intensity than the beatific vision of Heaven, when there will be vision without defect and love without weariness.[26]

Augustine also speaks of love in the ecstatic experience, as the "soul's weight" and the "soul's foot," the path by which the soul ascends to God and descends again. God's love enters the heart and gives new knowledge and new longing, which functions as a new sort of eye of the soul, a new way of seeing which is inspired by God.[27] Love is the "glue" which binds us to that which we desire, as charity binds us to God. Augustine notes, "the more ardently we love God, the more certainly and calmly do we see him."[28] Thus, the aspects of ecstasy affect each other—the love of God brings the vision of God, and eventually an identity which is bound up with God through charity.

Some Catholic understandings of ecstasy interpret it as a sort of trance state, with the ecstatic incapable of voluntary movement. The theologians Antonio Royo and Jordan Aumann define ecstasy as "a gentle and progressive swooning which terminates in the complete alienation of the senses."[29] The person has a radiant expression, but cannot see or hear anything, and cannot be returned to his or her normal senses except by a superior. The body falls into ecstasy through the intensity of love of God.

Ecstasy may be of natural or supernatural origin, and it is the latter type that is most important for Catholic mystical Theology. Supernatural ecstasy has two important elements: intense attention to some religious subject (interior and invisible) and alienation of the sensible faculties (corporeal and visible).[30] It may also be called flight of the spirit, elevation of the spirit, rapture, and other terms (some writers equate these, and others distinguish between these). Such ecstatic states often begin in childhood and may range in frequency and duration.

Some forms of ecstasy emphasize visionary experience, often described as both blinding light and great darkness, and sometimes these include perception of the personal forms of the Trinity. Ecstatic experiences are often marked by emotional extremes of both joy and suffering. There is great bliss and love and supreme delight, but sometimes also a sense of guilt, unworthiness, sorrow over personal sin and the damned in Hell, and a reliving of Christ's agony. Biographies of Catholic saints may emphasize both of these aspects.

The person's identity may undergo different types of transformations through the union of love. As St. Gregory the Great describes a vision by St. Benedict, the soul is expanded:

> St. Benedict "saw a light which banished away the darkness of the night... upon this sight a marvellous strange thing followed... the whole world,

> gathered as it were together under one beam of the sun, was presented before his eyes." St. Gregory adds… "For by means of that supernatural light the *capacity of the inward soul is enlarged*… But albeit we say that the world was gathered together before his eyes, yet were not the heaven and earth drawn into any lesser room than they be of themselves, but the soul of the beholder was more *enlarged*… and therefore, in that light which appeared to his outward eyes, the inward light, which was in his soul, ravished the mind of the beholder to supernal things."[31]

Sometimes the transformations of identity are described in almost a Vedantic style, emphasizing a union with God, as we can see in this description by St. Alphonsus Liguori of the spiritual marriage of the soul and God:

> In the spiritual marriage, the soul is transformed into God and becomes *one* with him, just as a vessel of water, when poured into the sea, is then *one* with it.[32]

A similar metaphor is used by St. Teresa of Avila:

> But spiritual marriage is like rain falling from the sky into a river or stream, becoming *one and the same* liquid, so that the river water and the rain cannot be divided; or it resembles a streamlet flowing into the ocean which cannot afterwards be dissevered from it. This marriage may also be likened to a room into which a bright light enters through two windows, though it passes through the two, *the light is one*…[33]

Catholic theologians Royo and Aumann describe ecstasy as due to the operation of love by the Holy Spirit, originating in infused or deep contemplation, in order to sanctify the soul. They divide ecstasy into three types: supernatural (elevation of the soul to God, and detachment from the sensory world), prophetic (involving illumination of the intellect), and mystical (the union of the soul with God).[34] These forms may be gentle and delightful, or violent and painful, but they are all ecstatic. Such ecstasy also brings a variety of physical symptoms: coldness, no apparent breathing or circulation, the inability to feel blows, wounds, or burning, to see bright lights or hear loud noises.[35]

These authors note that, though descriptions of ecstatic and mystical states often sound as if there is an absolute union with God, this is theologically unacceptable in Catholicism. Despite the language of mystical union (and the previous descriptions by St. Alphonsus Liguori and St. Teresa of Avila are good examples), Catholic Theology understands that these states do not involve any ontological change in the person. Rather, there is a transformation of the qualities of intellect and will, which had occurred in an unconscious manner at baptism, and now becomes conscious. There is mutual surrender and intimacy instead of union.[36]

While ecstasy is usually associated with the contemplative and ascetic life, there is also a tradition of ecstatic activism, where ecstatics write books, found orders, deal with land and money and politics, and in some cases marry and raise families. St. Augustine emphasized both the active and contemplative styles of life, the contemplative life as more favorable to ecstasy, and the active life as one's moral obligation. In the sixteenth century, St. Teresa of Avila was famous as both an ecstatic and an activist, as were many medieval visionaries. Here, we shall look at a contemporary of St. Teresa, best known for his writings on the Spiritual Marriage and the Dark Night of the Soul.

St. John of the Cross (1542–1591) was famous for his understanding of ecstatic states. He was a member of the Discalced Carmelites in Spain, and a Mystical Doctor of the Church. He had a special interest in the spiritual marriage, which he understood to be described *par excellence* in the Bible's "Song of Songs." He believed that when the human spirit reached its highest point, the spiritual marriage with Christ, the love of God for the person would meet the love of the person for God, and human corruption would fall away. Divine love then grows like a fire on a log, purifying desires which rise like smoke, and the material side crumbles to charcoal and falls away. This ecstatic love grows from a temporary state to a permanent one, and in the Spiritual Marriage the person is transformed.

For John, the person is composed of body, soul, and spirit. The soul consists of the external bodily senses (oriented toward the world of creatures), the natural desires and appetites, the natural passions and emotions (joy, hope, sorrow, and fear), and the interior bodily senses of fantasy and imagination. All of these are a part of the sensitive or lower nature.

The spirit consists of the faculties of intellect, memory, and will, and a deeper area which John calls the apex of spirit or substance of the soul. The intellect reasons and understands, receiving both natural and supernatural information. When purified, it can perceive spiritual light. The memory holds ideas and images and performs both passive storage and active recall. The will holds the power of love, and it is the basis of the person's strength and energy. It must be cleansed of the passions in order to find joy in God. The apex of the spirit is where ecstatic vision and transformation take place.

The ecstasy of the Spiritual Marriage is the climax of a process, the path of faith through darkness or the Dark Night of the Soul. The "night of departure" sends the person from identification with the "house of all appetites," the mortal body, to a path of darkness and eventually to union with God through love. The person is depressed, as John says, "He [God] leaves the intellect in darkness, the will in aridity, the memory in emptiness, and the affections in supreme affliction, bitterness and anguish, by depriving the soul of the feeling and satisfaction it previously obtained from spiritual blessings."[37]

This suffering cleanses the senses and desires and occurs during the Active Nights of Sense and Spirit (active in the sense of voluntarily chosen), accompanied by meditation and mortification. There are also Passive Nights of Sense and Spirit, where purification comes through God's will rather than human action. Nearing the end of this process, the soul undergoes a spiritual death:

> Since the divine extreme strikes in order to renew the soul and divinize it (by stripping it of the habitual affections and properties of the old man to which it is strongly united, attached and conformed), it so disentangles and dissolves the spiritual substance- absorbing it in a profound darkness- that the soul at the sight of its miseries feels that it is melting away and being undone by a cruel spiritual death; it feels as if it were swallowed by a beast and being digested in the dark belly, and it suffers an anguish comparable to Jonas' when in the belly of the whale (Jon 2:1:3). It is fitting that the soul be in this sepulchre of dark death in order than it attain the spiritual resurrection for which it hopes.[38]

This is followed by a vision of divine beauty:

> As the sun shining brightly upon the sea lights up great depths and caverns and reveals pearls and rich veins of gold and other minerals, so the Bridegroom, the divine sun, in turning to the bride so reveals her riches that the angels marvel and utter those words of the Canticle: Who is she that comes forth like the morning rising, beautiful as the moon, resplendent as the sun, terrible as the armies set in array (Ct 6:9).[39]

This vision brings love, and the bright flame of love fills the person with joy and light. It brings about the spiritual marriage: "… she [the soul] knows that at the instant she sees this beauty she will be carried away by it, and absorbed in this very beauty, and made beautiful like this beauty itself, and enriched and provided for like this very beauty."[40] It is a transformation of identity:

> The spiritual marriage… is a total transformation in the Beloved in which each surrenders the entire possession of the self to the other with a certain consummation of the union of love. The soul thereby becomes divine, becomes God through participation, insofar as is possible in this life.[41]

John calls the transformed self the "new man," whose center is God, and who has regained the Image of God. This is a long-term ecstatic state, in which the "habitual union of love" brings joy and bliss. This ecstatic state is best expressed in service, when every action is love.

Catholic mystical Theology examines many less prominent aspects of ecstasy as well: how to distinguish it from disease, insanity, and demonic intervention, how to analyze revelations and visions, the difficulties and trials of contemplatives and ecstatics, and the role of rare events which accompany ecstatic states, such as stigmata and levitation.

The Catholic Church and the Eastern Orthodox Church were united as mainstream Christianity until the Great Schism of 1054. At that point, they split into two different churches with different languages, rituals, and institutional structure. However, the disagreements between Roman and Byzantine forms of Christianity had been growing long before the official split, and one major area of dissension was the ultimate goal of Christianity for its members.

In the Orthodox tradition, the contemplative aspect came to be emphasized, and rational argument and systematic Theology never developed in the Eastern Church as it did in the Western one.[42] *Theologia*, or Theology, is a part of *theoria*, contemplation, as not rational argument but as a part of religious experience.

In the Eastern Orthodox tradition, the most important and respected form of ecstatic experience is *theosis* or deification. It is an experience of union with God's radiant love and Energies, though not with the individual persons of the Trinity, known as the Essence of God. This distinction between the Essence and Energies of God had been made by such Orthodox theologians as Gregory of Nyssa, Basil the Great, Gregory of

Nazianzus, John of Damascus, and Pseudo-Dionysius the Areopagite.[43] As Basil the Great states,

> It is by His Energies that we can say that we know our God; we do not assert that we can come near to the Essence itself, for His Energies descend to us, but His Essence remains unapproachable.[44]

The Orthodox belief in deification comes from their interpretation of several lines in the Bible. Psalms 82:6 is often translated, "I say you are gods" and it is quoted in the New Testament by Jesus (John 10:34-35). Another quote comes from 2 Peter 1:4, which states that through God's promise men "might participate in the divine nature." These are taken to mean not imitation or similarity, but rather that a person is transformed by divine grace in the process of *theosis* or *theopoiesus.*

Deification or *theosis* is strongly associated with spiritual light, of the sort that Moses saw at the burning bush, and in the darkness of Mr. Sinai, which transfigured him so that his face shone (Exodus 34:30). Paul's vision of light on the road to Damascus, and the disciples' vision of Jesus' transfiguration in light on Mr. Tabor (Matt 17:1-8) are also cited. The experiences of significant Biblical figures are important as evidence for deification.

Orthodox theologians echoed the emphasis on deification—we see this in Irenaeus' statement "If the Word is made man, it is that men might become gods," and Athanasius' slightly more pithy version, "God became man so that men might become gods" and Basil the Great's phrase, "The highest of all things to be desired [is] to become God."[45] Origen wrote of the spirit being "deified by that which it contemplates"[46] and Gregory of Nyssa noted, "God united Himself to our nature in order that our nature might be made divine through union with God."[47] Many respected orthodox theologians have written on this topic. In *theosis*, the person is said to undergo a transformation, participation, transmutation, union, intermingling, and elevation. There are many metaphors of marriage, rebirth, and re-creation.

As Georgios Mantzarides writes, deification or *theosis* is the ultimate goal of human existence, and the religious ideal of Orthodoxy.[48] It is regaining the Image of God lost in the Fall, a resurrection of the soul, preceding the resurrection of the body which will take place on the Last Day. It should be accompanied by dispassion or *apatheia*, a freeing the self from passion and gaining peace. As the early Greek Christians understood it, *apatheia* restored the person's spiritual balance and powers. Evagrius Ponticus called love the grandchild of *apatheia*.[49]

The *Philokalia*, a collection of writings by Orthodox contemplatives, gives many examples of these ecstatic states. As St. Makarios of Egypt states:

> The seeming joys of the world are not to be compared with the experience of divine grace in the soul. Those who share in this grace are sometimes filled with an inexpressible and nameless joy and exultation, as if they were at some royal banquet; sometimes they feel like bride and bridegroom delighting together spiritually, and sometimes like bodiless angels, since the body has become so weightless and light that it seems that they are not clothed with it. Sometimes it seems that they are in some realm greatly rejoicing and drunk with the inexpressible drunkenness of the mysteries of the Spirt, and then at other times they are full of grief, weeping and lamenting as they intercede for man's salvation. For, burning with the divine love of the spirit for all men, they take into themselves the grief of all Adam; and sometimes they are kindled by the Spirit with such untold love and delight that, were it at all possible, they would clasp everyone to their breast...[50]

This perfection of vision is understood to be legitimated by Christ's incarnation, which blessed humanity by the descent of the Logos or Word. As God came down and became man, man could ascend and become God. It can occur through the ecstatic union of the human spirit with the Energies of the Holy Spirit and is the highest and most valuable state that a person can attain.

ISLAM

Mainstream forms of Islam are not generally associated with ecstasy—it is a religion of obedience to God and his commands, of reading and interpreting the Qur'an and the Hadith, of following tradition. However, there are many types of Islam, and some types of ecstasy are allowed and respected. Here, we shall examine Shia and Mevlevi understandings of ecstatic states.

The major types of Islam in the world today are Sunni and Shia. The majority Sunni tradition follows the *sunnah* or traditions of Muhammad, with the five ritual pillars of Islam as the most important events. The Shia tradition, found primarily in Iran, is more apocalyptic and has a greater focus on tragedy. Sorrow and mourning are important, for they believe that the Sunni caliphs killed the rightful descendants of Muhammad, the *imams* or true leaders who were in the bloodline of Muhammad. There are also minority forms of Islam in the Sufi traditions, mystical schools which grew up in the Middle East and northern Africa.

We see examples of ecstasy in both orthodox and mystical schools of Islam. Muslim ecstasy tends to be based on important stories: on the description of Muhammad's ascent to heaven in the Night Journey or *miraj* in the Qur'an and Hadith, on the tragic death of Muhammad's descendant Husayn while he was on pilgrimage, and the passionate love of Layla and Majnun. The *miraj* has acted as a role model for Sufis who visit the throne of Allah, who see the angels and prophets as Muhammad did, and who wish to claim a precedent in a sacred text. The death of Husayn has acted as the historical basis for the yearly ritual of Ashura during Muharram, when Shia devotees torture themselves to share the pain of the victimized Husayn and his family.

The popular Arabic story of Layla and Majnun is the story of ideal divine love, based on a human model. The bedouin boy Qays (later known as Majnun, the mad one) loved the sheltered and beautiful Layla, his childhood friend, whose parents would not allow him to marry her. He went insane due to his separation from her. He saw her as perfect beauty and was so in love with her that he came to see her everywhere. He followed an interior journey in the wilderness and eventually he found Layla within his own heart, and all around him. Like God, she was everywhere in the world and also present within himself. The image of Majnun lost in contemplation of Layla acts as a model for the mystical lover, showing the transformation of human love into divine love.

Ecstasy in Islam is not always a joyful event. As the Muslim theologian and mystic Shaikh Shahabu-d-Din Suhrawardi phrases it in his thirteenth century *A Dervish Textbook ('Awarifu-l-Ma'arif)*, ecstasy or *wajd* is an event that comes from God and turns the heart to "great grief, or to great joy."[51]

In the Islam of the Shia Ali or the followers of Ali and his household (commonly known as *al-shi'a*), the most intense ecstatic state is not joy, but rather grief. It occurs at the yearly commemoration of the torture and death of Husayn, the grandson of Muhammad. This event, which happened on Ashura day during the month of Muharram, was of cosmic significance. The death of the rightful heir to the caliphate, a noble person in Muhammad's bloodline, represents the death of innocence, the injustice of the world, and life in the House of Sorrows which calls forth the coming savior, or Mahdi. Theologically, Husayn was understood to be the innocent sacrifice who would save the world, the lamb whose death transforms the universe. There are many parallels between the death of Husayn for the Shia, and the death of Christ for the early Christians. As Mahmoud Ayoub states:

> The martyrdom of *Imam* Husayn has been regarded by the *Shi'i* community as a cosmic event around which the entire history of the world, prior as well as subsequent to it, revolves… Husayn's death may be regarded as a redemptive act… through the participation of the faithful in the sorrows of the *Imam* and his beloved family. It will be seen, moreover, that not only mankind, but all creation as well, is called upon to participate in this tragic event.[52]

By his death, which was understood to be voluntary, Husayn transformed suffering itself from a negative and evil power to a road to salvation. This was due to human faith and divine mercy.[53]

One shares in the grief and sorrow of Husayn by mourning for him, especially each year on Ashura day during the month of Muharram. There are special mourning rituals intended to generate ecstatic grief. These include seeing the passion plays of Husayn's death or *ta'ziahs*, going on pilgrimage to Karbala, and torturing the body to physically imitate the pain experienced by Husayn and his family. There are long recitations known as *majlis*, which evoke sorrow and terror in the listeners, and are also considered to be purifying—the tears are said to cleanse the brain, making it easier to see the right path.

The goal of the annual Ashura rituals is ecstatic sorrow. This is grief which ceases to be individual and is understood to be a cosmic emotion. The entire universe mourns for Husayn, humans and animals and nature itself. The prophets knew that Husayn must die, and Adam and Noah and Ibrahim and the rest mourned long before Husayn was born. It is said that the sky shed tears of blood for Husayn for forty days, and even wild beasts roamed in anxiety in the jungles and forests. The sun darkened at noon, and stars collided with each other. Nature mourned Husayn in the past and continues to mourn him. All of the visible and invisible creatures of Allah mourned for Husayn, the *jinn* and men and angels. Seventy thousand angels will continue to weep for him until the Day of Judgment. The lower angels had to cool the burning fire of their sorrow with their wings, to prevent the heat from creating floods. As the sixteenth-century treatise *The Garden of Martyrs* states in a lamentation poem:

> Earth and heaven weep at the death of Husain;
> from the Throne on high to the dirt far below, all beings weep.
> Fish in the ocean depths, birds in the sky's upper heights:
> all weep in mourning for the King of Karbala.[54]

Such mourning brings unity with mankind and nature on earth, and salvation in the afterlife. As the eighth *imam* stated, "He who will remember our sufferings and afflictions and will weep and wail for us will be with us in the same (special) ward of Paradise."[55] Because weeping is the highest devotion and worship willed by Allah, the Shia who mourn go to the highest heaven, to be with the prophet and his family.

It is shared emotion which determines one's locale in paradise. As Imam Raza said to Bin Shabeeb: "If you wish to be with us in the most distinguished ward of Paradise, then feel our grief as your own grief and our joy as your own joy."[56] Such emotion is always possible because, as Imam Riza phrases it, heaven is eternally Ashura day, when Imam Husayn was killed.[57] Ali and Fatimah are eternally clad in black robes, sorrowing and weeping in paradise, until the end-times.

Ecstatic sorrow is total immersion in cosmic grief, and the person joins "the universal chorus of mourners." Tragedy is happening now, and every Muharram becomes the month of death, and every Ashura is the day of Husayn's martyrdom.[58] Those who participate in this mourning are marked by Husayn's blood and considered to be a part of his family. At the final resurrection, they will go to the area of paradise meant for the *ahl al-bayt* and will be considered to be a member of Muhammad's family. This sorrow is publically shared on Ashura day, when crowds of worshippers will whip themselves until blood flows, staining their white clothes red, to visibly show their sorrow for Husayn. Religious ecstasy in Shia Islam is not shared joy but shared sorrow. In its emphasis upon the universality of sorrow, it is reminiscent of Theravada Buddhism.

A very different understanding of ecstasy is found in the Mevlevi order of Sufism, based on the teachings of Jalaluddin Rumi. He was a thirteenth-century mystic known for his poems and books, especially the *Mathnawi*, often called the *Mathnawi Sharif* (the Holy Mathnawi) or "the Qur'an in Persian."[59] The order was institutionalized by his son, Sultan Walad, after Rumi's death. It was named after Rumi's title "Maulana" or "Mevlana," meaning "our Master," and called the Maulawi order in Persian, and Mevlevi order in Turkish. The group is popularly known today as the Whirling Dervishes. This is because their major ritual, the *sama* or sacred dance, involves the Sufis whirling around the room, as angels dance before the Throne of God, as bees and flowers dance in joy, and as the stars dance around the Pole Star.

According to Rumi's disciple and biographer Aflaki, Jalaluddin was a visionary as a child, seeing angels by the age of five years, and going on spiritual journeys with men in green cloaks.[60] He was a religious scholar and teacher, who started going into public ecstatic states when he met a dervish named Shamsuddin of Tabriz.[61] Jalaluddin understood him to represent perfection in human form. After Shams disappeared, the passions he evoked in Rumi came to be expressed in his poetry and dance. According to one story, Rumi's dance began when he heard a goldsmith's hammering of gold leaf, and he felt it to be the rhythm of the universe.

Rumi "engaged in passionate excitement and ecstasy at the *sama*" with his friends and companions.[62] He would see spirit beings (*ruhaniyan*) as he danced at the *sama*, and sometimes he would strip off his clothes and dance naked, "overcome by intense ecstatic emotion," and he would give his clothes to other people there.[63] Sometimes he would shout and cry until dawn in the streets, and his glance could bring people into ecstatic states.[64]

The *sama* was intended to generate ecstatic love through music and dance. Ecstasy is passionate emotion, as Rumi stated, "Ecstasy is the unrest of the spirit due to the overwhelming power of yearning."[65] But God is also ecstasy, according to Rumi:

> … God is entirely ecstatic delight (*dhowq*)… I am that ecstatic delight, and I am wholly immersed in that delight, whereas the [physical] delight of mankind is the opposite of that delight because 'Faith is entirely ecstatic delight and passion.'[66]

Music and dance motivate ecstasy, for "the House of Love is made completely of music, of verses, and songs."[67] The music of the *sama* is the sound of the doors of paradise opening. *Sama* is the nourishment of the soul, which brings the dead to life, and treads the grapes of life into spiritual wine. The *sama* does not represent an individual ecstasy, but rather a group one, where the lovers of God dance and sing together, and they sense his presence among them.

Rumi was a poet rather than a systematic theologian, but the ideas in his poems and stories reflect Sufi ideas of religious anthropology and the nature of God. Mankind is composed of three major parts: the lower soul, the heart, and the higher soul or spirit. The lower soul, or *nafs*, in its natural state brings temptations to lust and power. The heart, or *qalb*, brings faith and love of others. The *ruh*, or spirit, brings the love of God. It is the heart which is the intermediary (*barzakh*) between the soul and spirit, and also between the worlds of creation (*'Alam-i-Khalq*) and the divine world of command (*'Alam-i'Amr*), the worlds of mankind and of God.[68] There is also the intellect, which Rumi did not emphasize, and he referred to it as a lame donkey carrying books. The spirit seeks God, through the states of *fana* and *baqa*.

The state of *fana* or annihilation means that the person's individuality is destroyed before God, in order to take on God's attributes. It is a nullification of the ego, compared to black light, in which all things are absorbed and disappear.[69] In silence the person becomes empty, as he was before he existed as an individual. *Fana* often appears as madness, for the identification with God is not integrated, and it is unstable. The person is immersed in love of God (*ishq*), eternally seeking and burning for God. It leads to the more balanced state of *baqa*. In this state, the person becomes transparent to God's light, like Moses' burning bush. This is the light of the faithful, which extinguishes the fires of Hell, and the person's heart is like a lamp shining with divine light. In the folk tradition, it gives such abilities as telepathy, youth and strength, and the ability to purify and motivate others.

Rumi's focus was the ecstasy which linked the stages of *fana* and *baqa*, the divine intoxication which destroys the ego and gives glimpses of God. He had ecstatic raptures unpredictably and spoke of his flights of the soul in the claws of the moon-bird, and having the house of the heart illuminated. He spoke of drinking and becoming intoxicated by the wine of *kibriya*, the sight of God's majesty and grandeur. Ecstasy, or *wajd*, comes from the sight of this divine glory. In this vision, the person becomes like cloth dyed in a vat, radiant white from all colors merging together, or like a bar of iron which heats up in the fire and becomes like fire. Or more poetically, the ecstatic sees the sun dancing and clapping his hands in the sky, the atoms playing like lovers, the fountains being intoxicated and the roses smiling. Such ecstatic visions are a major aspect of spiritual growth in Sufism.

Hinduism

There are many ecstatic traditions in Hinduism, some described in Chapter Eight of this book. As an example here, I will include Bengali Shaktism or goddess worship, with the eighteenth-century visionary mystic Ramprasad Sen as its most famous poet.

In Bengali Shaktism, the goddess may be worshipped as a powerful force or life and death, and a tantric and yogic goddess who gives supernatural knowledge. The goddess may also be understood as a loving mother who saves her devotees from painful rebirth and gives them protection from harm as well as entrance into her heaven. This is Shakta devotion, or Kali *bhakti*. This is the type of Shaktism which was emphasized and popularized by Ramprasad Sen in his poetry, with Kali Ma, the universal mother or innocent girl, who is sometimes frightening on the outside, but inwardly loving and compassionate. Shakta *bhakti* also emphasizes that intense love of the deity is more valuable than simple obedience, thus showing the influence of Vaishṇava ideas of emotional bhakti.

The goddess has a variety of roles—mother, child, friend, lover. She expects passionate devotion and dependence from her worshippers, who are most often described as her children. The most intense love is not romantic love, but rather parent/child love. Some devotees begin with the emotion of fear toward her terrible forms, but they end up loving her when they see the sweetness within her. The devotee loves extravagantly; he (or she) is unworthy, pleading, and swings between elation and depression, loving the world because the goddess is present there or hating it because she has not appeared. The ideal attitude is total dependence, with love evoking divine vision or *darsan*, in which goddess and devotee meet and recognize each other.

Ramprasad was known for his ecstatic visions of the goddess Kali, which he described in poems and songs. Bengali Shaktism emphasizes devotion and love, with acceptance of both the loving and angry aspects of the goddess. Ramprasad famously worshiped the goddess at midnight on a seat of five skulls, a classical tantric practice, and sat in meditation in a *pancavati* grove containing five sacred trees. According to one story, he saw his statue of the goddess Kali as alive, and when his wife came over and perceived that the Kali statue was alive, she fainted. It is said that he then worshiped his wife with devotional songs, and they both stayed in trance all during the night (a later story has the Bengali *siddha* Ramakrishna Paramahamsa worshiping his wife Sarada Devi in a similar way).

According to another story, Ramprasad performed the corpse ritual (*sava-sadhana*) and had a vision of Kali in her form as Adyashakti Mahamaya. Ramprasad later retreated from the world and from his family and spent his time in meditation. He is said to have died of love on Kali Puja, as the Kali statue was being immersed in the Ganges River, with the vision of the goddess before his eyes. His poetry is still popular in West Bengal today, as a major part of *Shakta padabali* or poetry dedicated to the goddess. He is often called the most beloved of Bengali poets, and its greatest religious practitioner or *sadhaka*.

Ramprasad's poetry and songs (called *syama sangit* or songs to the dark goddess) focus on a single great goddess, who is most frequently called Kali, though sometimes she is also called Durga, Bhairavi, Sita, Uma, and Kalika. There are also songs of secret *sadhana* practices, and songs which emphasized the importance of passionate love which must be experienced, and cannot be found in books or philosophies. Devotion draws the goddess like a magnet attracts iron, and the poems include both dualistic and monistic imagery.

As examples of his work, the following poems are translations from the popular booklet *Ramprasadi Sangit* or "Songs of Ramprasad." Ramprasad was often called "love-mad" (insane due to his love of the goddess) by his friends and family, but for Ramprasad their opinions were not important. Here, he describes his ecstatic experience (Kali is traditionally believed to have red feet, which are considered to be beautiful by her devotees).

> O Ma Kali, wearing a garland of skulls
> What an experience (*bhava*) you have shown me!
> You taught me how to call you
> And at the moment I chanted "Ma"
> You drove me to ecstasy! (*mahabhava*)
> Ma Tara, please tell me the source (of your sweetness)
> Where did you get this name full of nectar?
> When worldly people look at me
> They call me mad from love

The members of my family
Hurl curses and insults at me.
But whatever people say, dark mother
My faith will not waver.
Let people say what they want
I will chant the name of Kali forever.
If you get rid of this illusory world
Insults and pride are unimportant.
I have made your red feet my goal.
I am no longer concerned with worldly opinions.[70]

The poems have imagery of death and salvation:

Because you love the burning ground
I have made a burning ground of my heart
So that you, dark goddess, can dance there forever.
I have no other desire left, O Mother
A funeral pyre is blazing in my heart.
Ashes from corpses are all around me, my Mother
In case you decide to come.
Prasad prays, "O Mother, at the hour of death
Keep your devotee at your feet.
Please come dancing with rhythmic steps
Let me see you when my eyes are closed."[71]

We may also include briefly the descriptions of ecstatic states in the Kularnava Tantra, a major Bengali Shakta text. This is a medieval Hindu text which goes into great detail on mantras, gurus, and visionary experiences. It describes two major types of ecstasy: the piercing or *vedha* of *diksha* (initiation), and the ecstatic joy or *ullasa* which occurs during ritual practice. The person who undergoes these states and reaches the highest or *divya* state is the ecstatic or divine madman. As the text states, "He faints in bliss, he trembles, is dizzy, and feels as if he is reborn. His limbs are stretched, his hair stands on end, he laughs and cries and stutters."[72] His touch brings bliss, his glance burns up sin. All actions become ritual ones—food is a sacrifice to the deity Bhairava, sexuality becomes cosmic union, and conversation becomes the chanting of hymns.

While in this state of *divyonmada* or ecstatic madness, the person is seen as equivalent to the God Shiva Bhairava, as in alchemy when iron is struck by mercury and becomes gold. The Kularnava Tantra says that he is beyond control by the body and the senses, he gazes outward but looks inward. The Kula tantric path is said to unite meditation and pleasure, yoga and *bhoga*. Mind and body are directed toward the God and goddess, resulting in ecstatic states.

Bengali Shaktism is still an important religious tradition in West Bengal, and the festivals to the goddess are the biggest and most elaborate festivals of the year.

Buddhism

Buddhist psychology does not accept the existence of an underlying soul or spirit. It is based on the idea of *anatman*, no *atman* or eternal aspect of the self. The deepest truth is *nirvana*, a state which is beyond suffering and individual identity. People are deluded into believing in the existence of an individual self because of the five *skandhas*, the components of bodily form, feelings, perceptions, impulses, and background consciousness. These are mistaken for an eternal soul, and the person is liberated when these "bundles" are understood to be collections of illusions. They become transparent to the person's deeper nature, which is the pure freedom of *nirvana*. Mental events are organized by type, length of existence, positive and negative aspects, and emotional associations or "feeling-tones." Ecstasy is rarely included as a separate mental event—it tends to be linked with insightful awareness and positive emotion.

In this section, we shall look at two types of Buddhist ecstasy. Pure Land Buddhism is the most widespread form of Buddhism in Japan, and it emphasizes salvation by Amida, the Buddha of the Land of Bliss. Tibetan Buddhism is a broad and complex set of traditions, with many subtypes. Here, we shall look at the yogic tradition of Naropa, from the Kagyu lineage, with an emphasis on the bliss of the Inner Fire practice, as well as initiation and deity yoga. These are examples of the wide range of ecstatic states found within the Buddhist tradition.

Pure Land Buddhism has several sects, of which the Jodo Shinshu or True Pure Land school is most popular. It was brought to Japan in the twelfth century by Honen, and he emphasized dependence on the Buddha (*tariki*) rather than independence and relying on the self (*jiriki*). Help comes from Amida, the Buddha of Boundless Light and Infinite

Life, the embodiment of compassion, a refuge and protector. Honen made Pure Land a separate sect, in which human imperfection requires divine mercy.

Pure Land was further organized by Shinran Shonin, a thirteenth-century Japanese monk who followed a vision which caused him to renounce monasticism and follow Honen. Shinran found human nature to be evil and sinful, based on selfishness and passion. All human actions were self-seeking, even the wish for liberation was a desire of a selfish ego. People were "defiled, evil and polluted, without a pure mind."[73] Thus, they must depend on a selfless Buddha for virtuous action.

While the earliest Buddhism of Siddhartha Gautama emphasized the non-existence of soul or deity, and the ultimate goal of *nirvana* which leaves suffering behind, over the years many understandings of these ideas developed and changed. The Mahayana tradition emphasized compassion over wisdom and Enlightenment and spoke of the importance of bodhisattvas, who dedicated their lives to helping others. Over time, paradise worlds of Buddhas and bodhisattvas were described in the Mahayana literature, as places of peace and joy, and as states of awareness. The most popular of these was the Land of Bliss, the paradise of Amida.

Amida's Western Paradise is a place of beauty. It is also called Sukhavati, the paradise of happiness, in the earlier Sanskrit texts, and Jodo in Japanese (from the Chinese *ching-t'u*). Some interpretations say that people are reborn there when they die; some speak of ecstatic visions of the Pure Land that people have while alive, and others equate the Pure Land with *nirvana*, saying that the Land of Bliss is here and now. It has been described in two major sutras.

According to the shorter Sukhavati Sutra, people who perceive the Land of Supreme Bliss are in states of unspeakable joy and happiness. The paradise is full of radiant light. There are pools with golden sand and fragrant lotuses the size of cartwheels. There are jeweled pavilions which ring with celestial music, rains of flowers, and rare and exquisite birds. The Buddha Amida is praised by other Buddhas with names like Magnificent Sunlight, Pure Beaming Light, and Splendor of Fame.[74]

According to the longer Sukhavati Sutra, when people see Amida, they become full of joy and cease to be subject to sorrow and affliction. There are jeweled trees, flowers of blazing colors, and musical breezes. The water there cleanses the heart and gives spontaneous bliss, and there are ten thousand kinds of music, making exquisite symphonies. There are untold millions of people in the Pure Land, and they dwell in buildings with beautiful tapestries, and they eat food by merely thinking of eating. The lotuses send out beams of radiant light, and people enter a state of bliss when they feel the winds of harmony.[75]

The texts or sutras describe visionary experience, as in the case of Ananda, who had visions of light and saw Amida in the sky. He states in the Longer Sutra that he saw the Pure Land, saw its inhabitants and heard Amida's voice.[76] The Pure Land of Supreme Bliss is a field full of the purity and power of Amida Buddha. Devotees experience "extraordinary worlds of visionary grandeur" which can be shared:

> The meditator can "open up" or "display" these visions for others, who then share in the meditator's experience, perceiving it as a sort of revelation.[77]

In this tradition, we have intense emotion, joy and bliss, and also visions, with imagery of radiant light. We see altered sensation in other ways, such as hearing celestial music and smelling the fragrant flowers of the paradise. The paradise itself is intended to bring spiritual insight and eventually liberation—thus a change in personal identity.

Amida Buddha, who had accumulated a vast store of merits or good karma during his many incarnations, established the Western Paradise. He saves all people, whether good or evil, if they call upon him. Amida has both unhindered light, which represents the quality of the mother, and the Name, which represents the father whose activity brings about salvation. Thus, Amida can be viewed as both mother and father. People gain Amida's merit, and all transfer of merit comes from Amida Buddha to mankind.[78] It is Amida who transforms the person into a Buddha, and the person is passive before his divine grace, and loses attachment to ego and body.[79] Amida accepts all, good and evil, his infinite love accepts and forgives everything if people are sincere in their faith. Chanting his name in the *nembutsu* prayer (*Namu Amida Butsu*) is a technique of salvation and a cry of joy, recognizing Amida's grace.

Concepts of ecstatic experience of Amida have changed over time. In its early Mahayana days, the Pure Land was a place of quiet meditation, free of temptation. The person would meet Amida at death and be reborn in his paradise. Amida's appearance at death depended on the faith of the devotee—he might appear in glorious form with attendants to dedicated believers; in illusory form to less dedicated ones; and in a dream to lukewarm ones. Thus, visionary ecstasy occurred at death. To help it along, there was a death ritual in which a painted screen with the figure of Amida on it was placed before the dying person, who would hold a string connecting him or her to the picture.[80]

However, such ecstasy could also be experience through visualization, which the person imagines the Pure Land in great detail, until the imaginary image has been transformed into a real perception. Some visualizations work partially, as in this one from "The Meditation upon the Buddha of Everlasting Life":

> When you have thus meditated upon the land, you should visualize the jeweled trees. Visualize the jeweled trees one by one; then meditate upon them in seven rows; each tree is eight thousand miles high; all the trees have flowers and leaves of the seven gems; and all of them are perfect... there are five hundred million flowery mansions, like the royal palaces of the highest gods. And these mansions are the homes where all the children of heaven dwell: each of the children wears five hundred million divine and magic gems as ornaments... He who has thus meditated is said to have seen the Land of Happiness with its jeweled trees and jeweled ground and jeweled waters, but still only roughly; yet whoever has seen this shall be exempt from the most heinous sins for innumerable eons; and when his body dies, surely he will be reborn in that land.[81]

Shinran added another understanding of ecstasy—one enters the Pure Land when one has faith in Amida Buddha and gains immediate salvation. This sudden faith in Amida is the ecstasy that is described in the texts. There is Pure Land now, and Buddhahood at death.[82] The Pure Land is both in and out of time, and there is no separation between the earth and the Pure Land. From this perspective, Amida Buddha is not an objective being, the Pure Land is not a geographical place, and one's presence in the Pure Land represents a way of understanding the Dharmakaya as compassion.[83] The Pure Land is also described as equivalent to *nirvana*, thus both Enlightenment and union with the Buddha.[84] Thus, ecstasy may be understood as both theistic and non-theistic. In both cases, Jodo Shinshu ecstasy leads to intense joy, altered perception of the world, and the loss of an individual self.

There are many schools and lineages of Vajrayana or Tibetan Buddhism, each with different spiritual practices, teachers, and patron deities. Here, we shall look at the teachings of Naropa, an eleventh-century Buddhist master in the Kagyu lineage of Vajrayana Buddhism, focusing on Naropa's Inner Fire or Tummo yoga practices.

This form of Buddhist ecstasy is based on the idea that each person has a physical and also a subtle body. Rather than speaking of levels of mind, in this literature we have levels of body. The subtle body has a central channel (as in the *sushumna* of the Hindu Kundalini yoga system), with four chakras or psychic centers along it. Each chakra is a center of *nadis*, spiritual nerves or channels of energy. They are supported by *prana*, the breath-energy or vital wind which also supports awareness.

As Lama Thubten Yeshe describes the practice (informed by the commentary of Je Tsongkhapa), the inner fire meditations let the yogi alter the energies of the subtle body and experience intense joy and deep wisdom. The practice beings four types of joy: joy, great joy, extraordinary joy, and simultaneously born joy.[85] The inner fire burns away illusions and attachments, and the mind "explodes with an intense awareness of non-duality."[86] In the bliss of the rainbow light which appears during the practice, all things become clear as crystal, revealing the infinite behind the finite. The bliss is intense awareness, like lightning in the sky, a clear light which embraces all reality.

It is produced by visualization of the subtle body, using mantras to activate the inner fire, and draws the energy of the breath (*prana*) into the central channel. The inner fire is ignited at various chakras, and the energy at the crown chakra melts and flows slowly down the central channel. As it flows down to the throat chakra, there is joy and bliss, and at the heart chakra, there is great joy. When it reaches the navel chakra, there is extraordinary joy, and at the secret chakra there is simultaneously born joy. As Yeshe notes, "This fourth joy is the peak bliss, superbliss. This simultaneously-born bliss totally unifies with non-duality and actually becomes the wisdom of emptiness, the experience of the clear light."[87]

The practitioner should then reverse the direction of the energy, bringing it up the central channel, which gives these forms of joy and bliss in the reverse order. When the energy returns to the crown chakra, the mind explodes with intense awareness of unity, and an even more intense bliss is experienced. As Yeshe says of this state,

> This bliss is not concrete, it has the nature of rainbow light. Everything becomes light when you unify bliss with nonduality. Everything assumes a transcendental appearance and becomes as clean-clear as crystal… everything becomes emptiness, the bliss becomes wisdom; it becomes clear light embracing the entire universal reality.[88]

Garma Chang notes that Tibetan Buddhist tantra has two major types of yoga—the path of liberation (mind yoga) and the path of skillfulness (energy yoga).[89] The first of these is observation of the Innate Mind, and the second involves rigorous and complex yoga practices, especially the arising and perfecting yogas. Chang places the yogas of Naropa in this latter category, of Anuttara or highest yoga. In the arising yoga, the body and world are understood to be mandalas or spiritual pathways. By the use of mantra and visualization, the person is able to perceive the subtle body, or identify with a particular Buddha. In the perfecting yoga, the self is dissolved into the light of the Dharmakaya and later projected outward to act as a Buddha.[90]

Ecstasy appears during many of these yogic practices, especially breath control (where colored *pranas* dissolve and circulate through the subtle body), and in the inner fire yoga. The blissful heat coagulates spiritual energy into drops (known as *bindu* or *thig le*) which melt and drop down, producing bliss. Such bliss may arise from visualization of energy in the central channel, or from the Great Bliss Chakra and its 32 rivers or *nadis*. More rarely, it may also come from visualized sex with a *dakini* (here as a female spiritual guide), performed in order to generate Tummo heat.[91]

Like Yeshe, Chang describes the four types of joy or bliss. There is Nadi bliss (delight at touch, which shows that many *nadis* are "tamed" or placed under the person's conscious control); Prana bliss (a momentary sense of pleasure in the body); Increased Thig Le bliss (a feeling of warmth and ecstasy in the body) and Melting Thig Le bliss (a feeling of intense sexual bliss permeating the body). The Melting bliss can be male (intense bliss and lust which is hard to control), female (intense bliss and lust easier to control) or neuter (weaker bliss and lust).[92]

The goal of these various bliss states is to merge the ecstasy with the Void or *nirvana*, and gain liberation. The Yoga of Bliss requires stabilization of these states, especially as these states become associated with the chakras. Thus, we have the First Bliss in the crown chakra (where crude distractions are subdued), the Second Bliss in the throat chakra (where subtle distractions are tamed), the Third Bliss or Beyond Bliss in the heart chakra (which eliminates all distractions) and the Fourth Bliss or Innate-Born Bliss in the navel chakra (which reveals the innate light).[93] The breath energies or *pranas* are stabilized in the chakras, and this allows the ecstatic states to be more controlled by the person.

Such practices are often accompanied by spontaneous ecstatic visions. These may include visions of light (such as thunder light and rainbow light), of the paradises of the Buddhas of Light, and of holy places on earth (mostly in India, where yogis and *dakinis* meet). There are also associated miraculous events, in which the yogi can radiate beams of light and fly to other worlds.[94] The Inner Fire or Tummo Yoga meditations lead ideally to *nirvana* or the Void.

Lama Yeshe also notes a few other types of ecstasy associated with the yogas of Naropa. He emphasizes that initiation is important (otherwise, the complexities of meditation will be like putting too much luggage on a yak), and initiation may lead to certain types of ecstasy. Initiation leads to identification with a deity and is accompanied by great bliss during that process. One example of an initiation ritual is the Vajrasattva initiation, in which the person visualizes the Buddha and his consort in bodies of light, sitting on a lotus. There is a descent of brilliant white nectar from the deities, like a waterfall of purifying light. This waterfall rushes through the person's spiritual body, cleansing it and giving it crystalline clarity.[95]

The practices of Tibetan Buddhist yoga involve "the cultivation of bliss." They clearly involve the three aspects of our definition of ecstasy. We see intense emotion in the states of extreme joy, and the various forms of bliss, which arise both during practice and when people enter the highest states. We see altered perception in the visions of spiritual light and Buddhas and their paradises, as well as in the mantras, mandalas, images of the subtle body and accompanying images which arise during meditation. We see altered identity in the identification of the practitioner with Buddhas, and in the merging of the person with the different levels of the Void. As Yeshe states;

> Because your entire deity body is one blazing inner fire and you are one with the inner fire, you can see the entire universal reality without any obstacles. Your intense bliss is aware of the wisdom of nonduality... You experience utter bliss: blissful chakras, blissful channels, blissful everything.[96]

Such practices are rarely performed today. This is partly because the Tibetan culture has been through such a disastrous recent history, and partly because of the difficulty in finding competent teachers.

However, we should note that descriptions and categorizations of ecstatic experience are widespread in Tibetan Buddhism, especially in the yogic tradition of highest attainment or *anuttara yoga*. For instance, the *Kalachakra Tantra* (an eleventh-century Indian tantric text used in Tibetan Buddhism) analyzes sixteen types of ultimate emptiness or *sunyata*.[97] It describes four types of Bliss (*ananda*), four types of Supreme Bliss (*paramananda*), four types of Extraordinary Bliss (*viramananda*), and four types of Innate Bliss (*sahajananda*).[98] The *Kalacakra Tantra* associates bliss with sexual experience; as Wallace states:

> In the Kalacakra tradition supreme imperishable bliss is defined as tranquility (*santa*) which pervades the elements of every sentient being's mind and body and of the entire inanimate world. Thus, the body of every sentient being is the abode of immutable bliss and contains the four bodies of the Buddha. In beings who are bound to transmigratory existence, the blissful nature of the Buddha's mind manifests in the form of sexual bliss, in which the mind, for a brief time, becomes non-dual and free of conceptualization.[99]

This glimpse of supreme bliss which some people have during sexuality is believed to motivate sexual desire, and thus attachment, mental affliction and bondage to the world. However, if used properly, it can under certain circumstances generate legitimate religious experience of ecstasy, bliss, and liberation.

These are some examples of ecstatic states and their interpretations from a variety of religious traditions. We shall now look at what Western academics and theologians have made of such states.

Notes

1. For a good example of this, see the biography of Vijayakrishna Goswamin in June McDaniel, *The Madness of the Saints: Ecstatic Religion in Bengal* (Chicago: University of Chicago Press, 1989), pp. 62–69.
2. June McDaniel, *Offering Flowers, Feeding Skulls: Popular Goddess Worship in West Bengal* (New York: Oxford University Press, 2004), p. 47.
3. Joseph Dan, *The Heart and the Fountain: An Anthology of Jewish Mystical Experiences* (New York: Oxford University Press, 2002), p. 12.
4. *The New English Bible with the Apocrypha*, Isaiah 6.1-3 (New York: Oxford University Press, 1971), p. 816.
5. New English Bible, ibid., Ezekiel 1.1-2, p. 1006.

6. Some other examples of prophetic call may be found in Amos 7-9, Zechariah 1-2, Samuel 3, and Moses' call in Exodus 3 and 4.
7. Gerhard Von Rad, *The Message of the Prophets* (New York: Harper and Row, 1965), p. 41.
8. Ibid., p. 42.
9. Abraham Heschel, *The Prophets*, vol. II (New York: Harper and Row, 1975), p. 115.
10. Ibid., p. 128.
11. Ibid., p. 125.
12. Max Weber, *Ancient Judaism*, cited in Heschel, ibid., p. 126.
13. Gershom G. Scholem, *Major Trends in Jewish Mysticism* (New York: Schocken Books, 1954), p. 40.
14. Ibid., pp. 123–124.
15. Ibid., pp. 131–133.
16. Abulafia, in Scholem, ibid., p. 134.
17. Ibid., p. 137.
18. Ibid., pp. 137–138.
19. Ibid., p. 140.
20. Abulafia, cited in Moshe Idel, *The Mystical Experience in Abraham Abulafia* (Albany: State University of New York Press, 1988), pp. 144–145.
21. Abulafia, in Idel, ibid., p. 75.
22. The numerical analysis of the Hebrew letters, or *gematria*, is based on the idea that Hebrew letters and numbers are identical, so words may be analyzed according to the numerical significance of the letters.
23. R. P. August Poulain, *The Graces of Interior Prayer: A Treatise on Mystical Theology*, trans. Leonora Yorke Smith (London: Kegan Paul, Trench, Trubner and Co., 1910), p. 54.
24. St. Augustine, *Literal Commentary on Genesis*, 12.12.25, cited in Bernard McGinn, *The Foundations of Mysticism*, vol. I (New York: Crossroad Publishing Company, 1992), p. 254.
25. St. Augustine, *Literal Commentary* 12, cited in McGinn, ibid., p. 254.
26. See McGinn, ibid., p. 255.
27. Ibid.
28. St. Augustine, *The Trinity*, 8.9.13, cited in McGinn, ibid., p. 259.
29. Antonio Royo O.P. and Jordan Aumann O.P. *The Theology of Christian Perfection* (Dubuque: The Priory Press, 1962), p. 551.
30. Poulain, ibid., p. 243.
31. From *Dialogues*, ed. Migne, cited in Poulain, ibid., p. 278.
32. St. Alphonsus Liguori, cited in Poulain, ibid., p. 288.
33. St. Teresa of Avila, *Interior Castle*, seventh mansion, cited in Poulain, ibid., pp. 293–294.

34. Royo and Aumann, ibid., p. 551.
35. Ibid., pp. 553–554.
36. Ibid., p. 557.
37. St. John of the Cross, *Dark Night of the Soul*, II.3.3, in *The Collected Works of Saint John of the Cross*, trans. Kavanaugh and Rodriguez (Washington, DC: Institute of Carmelite Studies Publications, 1979).
38. St. John of the Cross, *Dark Night of the Soul*, II.6.1.
39. St. John of the Cross, "Spiritual Canticle" 21:14, in Kavanaugh, ibid.
40. St. John of the Cross, "Spiritual Canticle" 11:10, in Kavanaugh, ibid.
41. St. John of the Cross, "Spiritual Canticle" 22:3, in Kavanaugh, ibid.
42. Daniel B. Clendenin, *Eastern Orthodox Christianity: A Western Perspective* (Grand Rapids: Baker Books, 1994), p. 53.
43. Vincent Rossi, "Presence, Participation, Performance: The Remembrance of God in the Early Hesychast Fathers," in *Paths to the Heart: Sufism and the Christian East*, ed. James S. Cutsingerm (Bloomington: World Wisdom, 2002), p. 74.
44. Clendenin, ibid., p. 61.
45. All three quotes are cited in Clendenin, ibid., p. 117.
46. Origen, "On First Principles" 3.6.3, cited in Clendenin, ibid., p. 127.
47. Gregory of Nyssa, *Oratio Catechetica* 25, cited in Clendenin, ibid., p, 128.
48. Georgios I. Mantzaridis, *The Deification of Man: St. Gregory Palamas and the Orthodox Tradition*, trans. Liadain Sherrard (Crestwood: St. Vladimir's Seminary Press, 1984), p. 12.
49. Evagrius Ponticus, *Practical Chapters* 53, cited in Mantzarides, ibid., p. 78.
50. St. Makarios of Egypt on Love, cited in *The Philokalia: The Complete Text Compiled by St. Nikodimos of the Holy Mountain and St. Makarios of Corinth*, vol. 3, trans. and ed. by G. E. H. Palmer, Philip Sherrard, and Kallistos Ware (London: Faber and Faber, 1984), p. 324.
51. Shaikh Shahabu-d-Din "Umar bin Muhammad-i-Sahrwardi," in *A Dervish Textbook from the 'Awariful-l-Ma'arif, Written in the Thirteenth Century*, trans. Lieut. Col. H. Wilberforce Clarke (London: Octagon Press, 1980), p. 82.
52. Mahmoud Ayoub, *Redemptive Suffering in Islam: A Study of the Devotional Aspects of "Ashura" in Twelver Shi'ism* (The Hague: Mouton Publishers, 1978), p. 141.
53. Ibid., p. 24.
54. David Pinault, *Horse of Karbala: Muslim Devotional Life in India* (New York: Palgrave, 2001), p. 164.
55. Syed Mohammed Ameed, *The Importance of Weeping and Wailing* (Karachi: Peermahomed Ebrahim Trust, 1974), p. 13.

56. Ibid., p, 27.
57. Ibid., p. 56.
58. Ayoub, ibid., p. 149.
59. John A. Subhan, *Sufism, Its Saints and Shrines: An Introduction to the Study of Sufism with Special Reference to India* (New York: Samuel Weiser, 1970), p. 35.
60. Shams al-Din Ahmad-E Aflaki, *The Feats of the Knowers of God (Manaqeb al-arefin)*, trans. John O'Kane (Leiden: Brill, 2002/1350), p. 55.
61. Annemarie Schimmel suggests that Shams was a *qalandar*, a wandering Sufi ascetic who is not bound by the traditional rules of Islam. He claimed to have reached the highest mystical state, the Pole of all Beloved. See Annemarie Schimmel, *The Triumphal Sun: A Study of the Works of Jalaloddin Rumi* (Albany: State University of New York Press, 1993), p. 20.
62. Aflaki, ibid., p. 104.
63. Ibid., p. 337.
64. Glances are powerful in this part of the world, especially with the belief in the power of the "evil eye." Aflaki gives a more dramatic example of its power: when a disciple gives water to a thirsty dervish, the dervish looks at him with his "wondrous glance" and the disciple promptly has an orgasm. See Aflaki, ibid., p. 267.
65. Aflaki, ibid., p. 244.
66. Ibid., p. 128.
67. Annemarie Schimmel, *The Mystical Dimensions of Islam* (Chapel Hill: University of North Carolina Press, 1975), p. 183.
68. Subhan, ibid., p. 61.
69. Schimmel, *Mystical Dimensions*, p. 144.
70. Ramprasad Sen, *Ramprasadi Sangit* (Calcutta: Rajendra Library, n.d.), p. 47. This is a booklet of poems in Bengali, sold at temples and book stalls in West Bengal, and at Shakta *pithas*, or goddess shrines, for use by priests, singers, and devotees.
71. Ibid., p. 46. A partial version of this poem was used in a film about the Thuggee sect of India, called "The Deceivers."
72. *Kularnava Tantra, Mula, Tika O Banganubadasaha*, ed. Upendrakumar Das (Calcutta: Nababharata Publishers), 1363 B.S. Sanskrit text, with Bengali translation and commentary, XIV, pp. 64–65.
73. Alfred Bloom, *Shinran's Gospel of Pure Grace* (Tucson: University of Arizona Press, 1977 [1965]), p. 33.
74. Luis O. Gomez, *Land of Bliss: The Paradise of the Buddha of Measureless Light, Sanskrit and Chinese Versions of the Sukhavativyuha Sutras* (Honolulu: University of Hawai'i Press, 1996), pp. 145–151.
75. Ibid., pp. 177–185.

76. Ibid., p. 216.
77. Ibid., p. 323.
78. Bloom, ibid., p. 49.
79. Ibid., p. 71.
80. Taitetsu Unno, *River of Fire, River of Water: An Introduction to the Pure Land Tradition of Shin Buddhism* (New York: Doubleday, 1998), p. 180.
81. Stephan Beyer, *The Buddhist Experience: Sources and Interpretations* (Belmont: Wadsworth Publishing, 1974), pp. 120–121.
82. Unno, ibid., p. 181.
83. Ibid., p. 179.
84. Bloom, ibid., p. 79.
85. Lama Thubten Yeshe, *The Bliss of Inner Fire: Heart Practice of the Six Yogas of Naropa* (Boston: Wisdom Publications, 1998), p. 159.
86. Ibid., p. 159.
87. Ibid., p. 160.
88. Ibid., p. 163.
89. Garma C. C. Chang, *The Six Yogas of Naropa and Teachings on Mahamudra* (Ithaca: Snaow Lion Publications, 1963), p. 14.
90. Ibid., p. 12.
91. Ibid., pp. 66–68.
92. Ibid., p. 69.
93. Ibid., pp. 70–72.
94. Ibid., pp. 79–80.
95. Yeshe, ibid., p. 55.
96. Ibid., pp. 140–141.
97. Vesna A. Wallace, *The Inner Kalacakratanta: A Buddhist Tantric View of the Individual* (Oxford: Oxford University Press, 2001), p. 93.
98. Wallace, ibid., p. 157.
99. Ibid., p. 177. We should note that in this tantra, the Buddha has more than the traditional three bodies. A fourth body is added, that of the Sahajakaya world, which is known through perfect awakening in a single moment.

CHAPTER 3

Attacks on Ecstasy: Pathologizing in Academia

Ecstasy in the modern West has been condemned in a number of ways and in a variety of disciplines in the academic world. In the area of philosophy, it has been portrayed as a symptom of ignorance, fanaticism, and superstition and more recently as a false belief associated with the claim of essentialism. In sociology, it has been associated with deprivation, genetic deficiency, and the generation of "crisis cults." In political science, it has been a sign of cultural weakness, a tool manipulated by tribal groups against powerful colonial interests. In anthropology, it has been interpreted as a strategy of marginalized people to gain social status and a primitive and magical precursor of rationality. From the perspective of psychoanalysis, ecstatic states have been interpreted as mania, psychosis, and symptoms of schizophrenia. In the field of History of Religions, ecstasy has been a symptom of traumas and power concerns, symbolizing cultural domination and competing values, and a potential cause of violence. From the medical perspective, it is classified as a symptom of psychiatric ailments, brain misfirings, epilepsy, and even calcium deficiency.

From all of these perspectives, ecstasy ranges from an error to a danger. While some of the arguments for a negative view of ecstasy have been left behind by history, the negative view itself has largely remained. Academic fields have shown both ethnocentrism and moralizing in their condemnations of ecstatic states. The writers who have historically been more sympathetic toward ecstatic states (such as Carl Jung and Mircea Eliade) have also been attacked with great force today, both at academic conferences and in college classrooms. Why has the field developed in this way?

J. McDaniel, *Lost Ecstasy*, Interdisciplinary Approaches to the Study of Mysticism, https://doi.org/10.1007/978-3-319-92771-8_3

This chapter gives some major examples of the critical scholarship on religious ecstasy, following both senses of the definition—critical as important and also critical as skeptical and faultfinding. I use the term pathologize here: "to regard or treat (someone or something) as psychologically abnormal and unhealthy."[1] In a broader usage, ecstasy has been a problem for a wide variety of academic disciplines in their understandings of culture and society, and the condemnation of ecstasy as "abnormal and unhealthy" comes from many perspectives.

Philosophy of Religion

Suspicion of ecstatic states was a major theme in the writing of European Enlightenment thinkers. Their views set the stage for the later condemnation of ecstatic states by the sciences. Two of the most well known of these philosophers were Immanuel Kant and David Hume. Kant made the miraculous and supernatural unacceptable to the educated classes of Europe, portraying these phenomena as unproved claims outside of mankind's possible sphere of understanding. According to Kant, the only acceptable proof of God's existence could be found in the human conscience and the sense of morality. As we could not know the divine world of "things as they are" (*noumena*), it became distant and invisible, a mere hypothetical, with no connection to human experience of the physical world of *phenomena*. The divine world was beyond the possibility of rational human apprehension. This perspective that there can be no unmediated cognition of reality has been the dominant view in Western philosophy since that time. "Supernatural communication" and "mystical illumination" were, for Kant, opposed to philosophy.

In Enlightenment thought, the divine world became simply imaginary, a projection of human desire, and a way to justify human power cravings and the oppression of others. States of ecstasy were equated with enthusiasm, the awareness of "God within" (*en-theos*) which bordered on insanity, and with fanaticism, as a corruption of reason. In his *Critique of Judgment*, Kant distinguishes between enthusiasm and fanaticism:

> If enthusiasm is comparable to madness, fanaticism is comparable to monomania.. In enthusiasm, regarded as an affection, the Imagination is without bridle; in fanaticism, regarded as an inveterate brooding passion, it is without rule. The first is a transitory accident which sometimes befalls the soundest Understanding; the second is a disease which unsettles it.[2]

For Kant, rather than representing an external threat to rationality, the "disease" of fanaticism appeared as one possible form of reason and a particularly bad one. In his view, the project of "critique," his examination of the limits of what we can know, had to continually wrestle with the mind's tendency to trespass over its own boundaries and to "rave with reason," believing it could "see the infinite." Kant understood fanaticism as a misuse of reason's powers, a transgression of its own limits. It went beyond what was possible for reason, for it was clear to him that ecstatic states that held religious truth were irrational and impossible.

David Hume described two forms of false religion, superstition and enthusiasm, and he considered both to be problems. Superstition was full of dread and terror, gloomy and malevolent, leading through fear to priestly authority, which he found to be untrustworthy and dangerous. It brought "blind and terrified credulity." He contrasted this with enthusiasm or ecstatic joy. Hume noted that enthusiasm starts out in a positive way, but leads to vanity and ignorance:

> But the mind of man is also subject to an unaccountable elevation and presumption, arising from prosperous success, from luxuriant health, from strong spirits, or from a bold and confident disposition. In such a state of mind, the imagination swells with great but confused conceptions, to which no sublunary beauties or enjoyments can correspond. Everything mortal and perishable vanishes as unworthy of attention. And a full range is given to the fancy in the invisible regions or world of spirits... Hence arise raptures, transports, and surprising flights of fancy; and confidence and presumption still increasing, these raptures, being altogether unaccountable, and seeming quite beyond the reach of our ordinary faculties, are attributed to the immediate inspiration of that Divine Being, who is the object of devotion. In a little time, the inspired person comes to regard himself as a distinguished favourite of the Divinity; and when this frenzy once takes place, which is the summit of enthusiasm, every whimsy is consecrated: Human reason, and even morality are rejected as fallacious guides: And the fanatic madman delivers himself over, blindly, and without reserve, to the supposed lapses of the spirit, and to inspiration from above.[3]

Unfortunately, enthusiasm could also create delusion and fanaticism:

> Enthusiasm being founded on strong spirits, and a presumptuous boldness of character, it naturally begets the most extreme resolutions; especially

> after it rises to that height as to inspire the deluded fanatic with the opinion of divine illuminations, and with a contempt for the common rules of reason, morality, and prudence.[4]

Enthusiasm is not quite as bad as superstition, but both can bring violence: "It is thus enthusiasm produces the most cruel disorders in human society..."[5]

Another influential Enlightenment philosopher was Voltaire, who criticized the religious passions in his article "Fanaticism" and the play "Mahomet the Prophet." He described the fanatic vividly as the enemy of the Enlightenment, with the philosopher as the personification of tolerance. He describes fanaticism as a political perversion of the religious spirit, a destructive and contagious group fixation on divine commands which will not rest until contradictory ideas are destroyed. For Voltaire, all fanatics "have the same bandage over their eyes." The state of enthusiasm involved emotion, shock, passion, frenzy, madness, fury, and rage. He notes in his *Philosophical Dictionary* that "Enthusiasm is precisely like wine: it can excite so much tumult in the blood vessels, and such violent vibrations in the nerves, that the reason is entirely destroyed."[6]

For these writers, only secularism could avoid the dangers of both fanaticism and superstition. Secularism became the most reasonable approach to life, and it also opposed the religious wars which had shown their destructive power in Europe. What came to be preferred was morality without any expectation of religious experience, as shown in Deism. This approach accepted a distant and disinterested God, who created the world at the beginning of time, but was no longer concerned with human affairs. This lack of divine involvement made ecstatic claims illegitimate, and led to a positivistic outlook in which humanistic ethics would guide humanity, and science came to be the best way of knowing truth. In Enlightenment and post-Enlightenment philosophical writing, ecstasy came to be associated with ignorance, superstition, the violence of religious wars, and eventually theological resistance to scientific thinking.

The condemnation of superstition, and religious experience or enthusiasm, has been an ongoing theme in the history of philosophy. Philosophy of religion has come to discourage the study of ecstasy in two major ways. One approach emphasized superstition; as there is nothing out there to study, ecstasy is a response to a false belief about a "given" religious realm or ultimate state that does not exist. This approach has been described as naturalist, secularist, and anti-supernaturalist. The other approach states that if such religious realms and ultimate states did exist, we still could not know them because we only know language and concepts, and we cannot have actual knowledge of objects. We perceive only through social constructions and presuppositions, while claims about non-human sources rely on pre-social or extra social contact with a "bare perceptual given"; this has been characterized by Roderick Chisholm and others as "the myth of the given." This anti-ecstatic approach is usually described as constructivist or contextualist. For both of these groups, the primary approach should be critical, especially of theological claims, and philosophers should not only avoid advocating a theological position, they should actively combat theological claims.

The neo-Kantian perspective of the constructivists has become the dominant perspective in the study of ecstatic states. Perhaps the most well-known spokesman for the constructivist position is Steven Katz. From this perspective, ecstatic and mystical experiences are only cultural and linguistic constructions dependent upon a web of definitions and assumptions, which are tied to specific beliefs and practices. They are social constructions that have been created in different ways at different times, and do not reflect a single, deeper truth. As he notes,

> The experience itself as well as the form in which it is reported is shaped by concepts which the mystic brings to, and which shape, his experience ... The forms of consciousness which the mystic brings to experience set structured and limiting parameters on what the experience will be.[7]

Ecstatic claims here are an error, a belief in a false projection of ideas. Other philosophers have followed this line of thought. Wayne Proudfoot has claimed that all religious experience is constituted by beliefs and concepts, and that there is too much variety to assume that there is any universal truth beyond them. There is no experience separate from its interpretation; therefore, different interpretations mean different sorts of experiences; an experience beyond or outside of such symbolism could not be described or remembered.[8]

Robert M. Gimello also takes the constructivist position, arguing that it is a mistake to consider any mystical experience as autonomous, as societal ideas totally determine the types of realization that people have.[9] Philosophical views range from denying any essence or religious goal present in ecstatic states, to the distant possibility of there being a religious goal which human beings can never realize. While Robert Forman has argued against the constructivist position, in favor of an ecstatic "consciousness without an object," his arguments have not swayed most of the field.

Because all ecstatic experience is defined as derivative of previous states and social learning, there is no room for innovation or new forms of ecstatic experience. The constructivist view thus cannot respond to spontaneous ecstatic states which do not fit into a preexisting religious system. It can only deal with successful mystics who have been accepted into religious systems and whose lives and works are recorded and interpreted by them. People who have ecstatic and mystical experiences which do not fit into a cultural mold are simply rejected—either their experiences are signs of insanity, or they are meaningless views of a recognizable religion.

Attempts to found religions and spiritual paths are often unsuccessful, and these visionaries and prophets fade into obscurity. Religious institutions rarely keep records of unsuccessful ecstatic claimants. But when they do have some success, and begin to attract followers and create an organization, ecstatics become founders of religions and social institutions and can be studied by the field of sociology.

Sociology of Religion

In the field of sociology, ecstasy may be a form of primitive magic, a result of deprivation, or a compensatory response to a lack of social power, prestige, and wealth. If one cannot get power from one's bloodline or land ownership, there is always direct contact with deities to give status and authority. People who have too little will seek to have too much: excessive emotion, religious passion, wild rituals, and loud prayers.

Some theories from the nineteenth and twentieth centuries emphasized the intellectual inferiority of ecstatic people, noting problems of bad genes, primitive thinking, degeneracy, delirium, and even a lack of evolution. Writers distinguished between advanced and primitive cultures, with ecstatic events occurring in cultures that are more childlike, exotic, and irrational. Ecstasy may thus be understood as the enemy of ethics and rationality, as it has been in philosophy, and an escape from social responsibility.

Two classical writers in the field of sociology of religion are Max Weber and Emile Durkheim. In his 1922 book *The Sociology of Religion*,

Max Weber finds the origin of religion in charisma, extraordinary powers which can affect the world. The magician is a person who is permanently endowed with charisma through the experience of ecstasy. The magician is a specialist in ecstasy, which is "a unique object of his cultivation," for he defines ecstasy as "the distinctive subjective condition that notably represents or mediates charisma." It may arise through orgies or through intoxication, as Weber notes, "Unlike the merely rational practice of wizardry, ecstasy occurs in a social form, the orgy, which is the primordial form of communal religious association."[10]

While laymen only experience ecstasy occasionally, as intoxication, magicians experience it more often. He notes that "professional magicians and warriors need permanent states of charisma, as well as acute ecstasies."[11] The "capacity for ecstasy" is found in both the hero and the magician, and it involves the breaking down of inhibitions, which leads to the concept of a soul that leaves the body in ecstatic states, in fainting, in dream, or in death. Ecstasy "may have the essential character of an acute mental aberration or possession, or… a chronically heightened idiosyncratic religious mood" which can be contemplative or active.[12] However, such orgiastic ecstasy became the enemy of prophets of ethical salvation, as it violated the ethical patterning of life that they preach. Weber notes that orgiastic ecstasy was a primary target of such prophets as Zoroaster and Moses.[13]

Weber claims that as the understanding of the God becomes more transcendental, there is a loss of potential for ecstatic self-deification. The goal becomes the acquisition of religious qualities that the God desires in mankind, and God-possession becomes impossible. Such states may only be simulated through the "hysteroid deadening" of bodies, making them anesthetic or cataleptic in ecstatic rituals.[14] Ecstasies may be sublimated, into the Catholic semi-erotic mysticism of the Middle Ages, into "orgiastic dervishism," and the "intellectualist ecstasy" of India.[15] In the modern world, such ecstatic states tend to be sublimated into economic and class concerns, following the values of early Rome. As Weber states, the ancient Greeks accepted ecstasy, especially through Dionysus, while the Romans rejected such ecstatic states, translating the Greek word *ekstasis* into the Roman *superstitio*, a dangerous and antisocial excess.[16] Rome rejected every type of ecstasy, turning the term *orgia* from Dionysic religious rites to the orgy, in popular understanding an excess of food, drink, and sexuality.

Weber noted that when Christianity became accepted in Rome it adopted this dismissive attitude toward ecstasy. The Roman Church consciously avoided irrational elements, from charismatic prophecy to new forms of church music. The Roman focus on dogma and ethics was incorporated into early Christianity in the condemnations of the church at Corinth and the Hellenistic Orient.[17] Ecstasy became a threat to social order, and prophets preached against it, because new prophecies could disturb the ultimate truth and values of the older prophecies. Weber's understanding of the origin of religious institutions is the "routinization of charismatic power," after which ecstatic charisma becomes exiled from religious practice. It is the chief task of institutionalized religion to substitute a controlled system of social order for the earlier chaos of ecstatic inspiration. Ecstasy is rejected, pathologized as the enemy of rationality and ethics.

In Weber's work on the Protestant ethic, he notes that the most urgent task of Calvinism was the destruction of spontaneous and impulsive enjoyment, recognizing that life was a burden and full of sin.[18] Emotionalism was only a distraction from God's wrath and human damnation, and the major function of religion was to bring order into conduct via constant self-control. Ecstasy violated this necessary order.

For Emile Durkheim, the origin of religious life was in the ecstatic state of "collective effervescence," which occurred when the entire tribe gathered together. He based his study on Australian aboriginal Corroboree celebrations, which involved "an extraordinary degree of exaltation," exciting participants to a frenzy.[19] As Durkheim understood it, the God which is experienced in such rituals is a personification of the clan, a totemic emblem or divinity which is really a symbolic expression of social experience. This ecstatic experience serves the function of uniting the group, giving salvation from the "utter colorlessness" of boring work like hunting and gathering. People gather in "transports of enthusiasm" running like madmen, screaming with heated passions free from control. Durkheim claims that such hyper-excitement is a "contagion" for "unformed minds," which enter into a "pseudo-delirium" and project religious ideas into the environment. As Durkheim notes, religious force is "projected outside of the consciousnesses that experience it, and objectified. To be objectified, it is fixed upon some object which then becomes sacred."[20] Thus, the collective and anonymous force of the clan was projected into totems as a form of primitive classification, which allowed the superimposition of religious ideas on the world around them. Such ecstatic states had the virtue of creating bonds which held the group together. However, they were a sort of primitive, savage contagion, unsuited to the modern world.

Both Weber and Durkheim based their analyses of ecstatic states on fieldwork, in other cultures. However, many sociologists of the early twentieth century focused on the USA, where Christian revivals had participants who displayed numerous ecstatic states. As sociology of religion had not yet developed into a separate field, they often used psychological and anthropological analyses. As Sean McCloud notes in his book *Religious Distinctions: Class and Classification in Religious Studies and American Religion*, many early sociologists held a negative view of religion. He writes that they believed that "certain religious beliefs and practices naturally attracted certain races and classes of people."[21] Depraved people were naturally attracted to depraved religions—those which were considered to be "primitive" and irrational, opposing the tide of rationality and secularism which were the markers for "advanced" people. Eugenicists and sociologists agreed on a list of degenerate religions: Roman Catholicism, sectarian and minority religions, fundamentalist religions, ecstatic religions, and new religions.[22]

One major criterion for depravity was a religion that was emotional and ecstatic, attracting "morons" and other "less favored" individuals. People attracted to such religions were believed to have genetic or mental problems. Intelligence meant rationality, and an overly emotional person was understood as unintelligent and otherwise inferior, and religions could create or worsen depraved and insane people. The field of sociology was incorporating concepts of evolution at that time, and as McCloud notes, it was the participants in "ecstatic and experiential-ly-based religious activities" like revivals and speaking in tongues who were consistently described as unevolved.[23] The assumption was that people evolve spiritually as well as physically, moving from the primitive, ignorant, and emotional to the rational, moral, ethical, and controlled. Spontaneity was dangerous, while control was socially valuable.

An early writer who exemplifies this view is Frederick Moran Davenport. In his 1910 book *Primitive Traits in Religious Revivals: A Study in Mental and Social Evolution*, he describes the same sort of scale of human mental development ranging from primitive to advanced. He was interested in ecstatic states in American religion and believed that qualities or traits of the primitive mind are those shown in religious revivals: suggestibility, emotionalism, imagination, imitativeness.

Such traits need to be controlled, or they motivate lynch mobs, emotional revivals, and revolutions. The imagination unlocks the floodgates of emotion, which may become "wild enthusiasm or demoniac frenzy," spread through a "contagion of feeling."[24] Davenport's main concern was the ecstatic states to which primitive minds are prone; he describes these as frenzy, passion, catalepsy, collapse, foaming at the mouth, convulsion, total loss of inhibition, and people being "mad with supernatural joy."[25] He is clearly repelled by such states, including the "singing ecstasy," emotional shouts and groans, extravagances and disorders, swooning, and child preaching.

Davenport associates these states with "infectious disease" and "morbid contagion," as well as "moral degeneration," "spiritual decay," and being in a hypnotic state.[26] Human genetic deficiencies were described as showing "cacogenic" or inferior stock and condemned by writers interested in eugenics.[27] Depraved and inferior groups were inclined toward ecstatic religious practices and were dangerous, for depraved religions create depraved members.[28] Religious ecstasy caused a process of evolutionary regression, from the civilized to the primitive.

More recently, we see writing on ecstatic states in the literature on new religious movements, which have included prophetic revelations by charismatic leaders and visions of the end-times by apocalyptic groups. In the 1960s and 1970s, critiques of ecstatic states dealt with religious cults and the "brainwashing" that occurred with altered states of consciousness. Rodney Stark and William Sims Bainbridge discuss the "psychopathology model" of cult formation and innovation, described by many social scientists. It is another deprivation model, based on response to personal and social crisis. This type of cult is founded by people suffering from various forms of mental illness who have a "novel vision" during psychotic episodes, and these hallucinations motivate a new understanding of the world and a strong commitment to these visions. If the founder with his particular type of mental illness is able to find others who suffer from the same problems, he may be able to form a cult group (a paranoid leader, for instance, would be most attractive to similarly paranoid followers). Such groups tend to develop in times of social crisis, and the ecstatic visions of the cult leader become legitimated by others, and he gains rewards from them.[29] Thus, ecstasy may follow the "deprivation model" of seeking power and wealth, or the "psychopathology mode" of cult formation.

Anthropology of Religion

In the nineteenth century, the evolutionary model was also important for the growing field of cultural anthropology. Most anthropologists of the time assumed that there was a clear distinction between "primitive" and "modern" religion, and it was the primitive religions that had the greatest focus on ecstatic states like visions and dreams. More modern approaches to the anthropology of religion emphasize the ways that religious beliefs and practices may reflect and influence political, economic, or social forces. Some modern anthropologists are also trained in psychotherapy and work in the areas of ethnopsychology and ethnoanthropology.

Older well-known writers in anthropology who studied ecstatic states include Claude Levy-Bruhl, who coined the term "participation mystique" to describe what he considered to be the prelogical stage of savages. Levy-Bruhl believed that tribal people lived in a world of magico-religious collective representations, and experienced the supernatural through invisible forces in daily life, following the "law of mystical participation."[30] He described an atmosphere of spirits, witches, and enchantments in which the dead threatened the living, and he tended to equate the supernatural and the mystical, both involving awareness of magic and ghosts. Ecstasy was their experience of the supernatural, especially perception of the dead. Levy-Bruhl noted that such a view violated the laws of causality, and it organized space and time in an irrational fashion.[31] He changed his views later in life, and his ideas on primitive religion have been largely discredited, as later anthropologists have argued against absolute distinctions between primitive and civilized minds. However, his writings were a popular attempt for his time to understand ecstatic states.

The most well-known modern anthropological writer on ecstatic trance is probably I. M. Lewis. In his book *Ecstatic Religion: An Anthropological Study of Spirit Possession and Shamanism*, he defines ecstasy in terms of encounter: "the seizure of man by divinity."[32] He describes the person in such states as filled with "transports of mystical exaltation in which man's whole being seems to fuse in a glorious communion with the divinity."[33] However, this encounter is not merely a divine blessing. Rather, possession gives the possessed person increased power and status in his or her community: The person gains "authority to act as a privileged channel of communication between man and the supernatural."[34] It is a path to social power and privilege, and for Lewis, this is its primary function. He is interested in the religious authority that is bestowed by ecstatic states, and the ways that surrendering individual will allows the person to gain increased social power and control.

For Lewis, religious ecstasy is a form of trance state, involving dissociation and lack of voluntary movement, and it is accompanied by involuntary and automatic acts. Such symptoms are also seen in hypnotic and mediumistic states and may involve visions or hallucinations.[35] He notes that trance may be attributed to a variety of sources, including gods and demons. The Catholic Church has traditionally claimed control of the interpretation of these states in the West, and trances have often been declared as heretical and caused by the Devil.[36] Besides ecstatic or mystical trance, we also see non-mystical forms of trance, such as tarantism or St. Vitus' dance (especially prominent in the fifteenth-century dancing craze associated with convulsions and apocalyptic visions, which swept through central Europe).

Lewis describes a shamanic trance state that "begins as an illness, or otherwise deeply disturbing experience, and ends in ecstasy."[37] However, for Lewis shamanism is primarily a social phenomenon, concerned with healing and alleviating distress rather than attaining ecstasy. He notes that many cultures recognize that such trance states may be real or simulated, contrasting "inauthentic or negative possession" with divine ecstasy, which is "authentic possession."

Lewis writes that "outbursts of ecstatic effervescence" may signal the rise or decline of religions.[38] Many religions start with ecstatic fervor, but they lose this over time, becoming more rigid and forming "ecclesiastical establishments which claim a secure monopoly of doctrinal knowledge." Such a "hardening of the spiritual arteries" restricts religious change and growth, giving human beings the religious authority that once belonged to the gods.[39] The focus becomes dogma rather than ecstatic inspiration—a structure that is more stable, more predictable, and more secure than ecstatic trance. This is why socially stable churches emphasize ritual rather than ecstasy, for "enthusiasm thrives on instability."[40] He notes,

> New faiths may announce their advent with a flourish of ecstatic revelations, but once they become securely established they have little time or tolerance for enthusiasm. For the religious enthusiast, with his direct claim to divine knowledge, is always a threat to the established order.[41]

For Lewis, ecstasy is an abnormal, erratic condition, whose major appeal is the social status which it may bestow under certain circumstances. The religious and mystical claims of the ecstatic individuals are secondary to its social motivations and results. Ecstatic trance is a social pathology, generally a result of deprivation, and Lewis describes "a widespread use of spirit possession by means of which women and other depressed categories exert mystical pressures upon their superiors in circumstances of deprivation and frustration, when few other sanctions are available to them."[42]

Some of the most interesting fieldwork on ecstasy is written by anthropologists who also have medical and psychoanalytic training. The psychiatrist William Sargant has done anthropological field research on trance in Africa and the Caribbean. For him, the ecstatic state is a result of increased suggestibility, the same sort of state that people enter when they fall under the spell of the demagogue or revivalist, priest, psychiatrist, and even the propagandist and the advertiser.[43] He defines ecstasy as "when man feels taken over by a force or being greater than himself"[44] and he finds ecstasy clearly visible in the state of possession trance. In the rituals he saw, possession was deliberately induced, to give the person "the most direct and immediate possible experience of a deity, by becoming the living vessel, and to enable him to act as a channel of communication between gods and spirits and their worshippers on earth."[45] In these possession states, there is intense excitement, and a second personality emerges which replaces the original one. Sargant describes frenzied emotion, primarily anger and fear, which leads to a collapse, which is followed by calm.[46]

Mystical ecstasy, the "supernormal" state of mind, results from such stress in Sargant's account. He claims that there is a drastic disorganization of higher nervous activity and abnormal phases of brain activity. When a certain point of stress is reached, the brain enters "paradoxical activity," in which the person's threat or preoccupation is internalized and the fears of outer events become experienced as inner events. The person may feel the God, spirit, or Devil within him, and the state is described in paradoxical terms, such as "brilliant darkness" or "painless pain." It may come from solitary meditation, or from revivalistic or orgiastic activity.[47]

Though these appear to the experiencer to be real events, they are actually imaginary. As Sargant states, "it is the brain of man and not his soul which is affected by mystical techniques," and the possessing deity or spirit will be identified differently based on varying religious backgrounds.[48] Nevertheless, the ecstatic state gives the person absolute certainty. Paradoxical brain activity becomes ultra-paradoxical, and intense emotional stress evokes intense ontological certainty. This intense, temporary brain excitement leads to nervous collapse and brain inhibition. It may be accompanied by a powerful and uncritical transference (or sense of love and devotion) and a relief from tension.[49]

This is also the pattern for sexual ecstasy, in which people are swept away by an overwhelming force, which they are powerless to control, feeling as if seized from the outside. Orgasm is like possession trance; people writhe, cry out, groan, tremble, and are blind and deaf to everything around them. They may have a sense of union or mingled identity, before the intense state ends in collapse.[50] Sexuality for Sargant follows the same pattern as possession trance, and sexual ritual can be used to induce ecstatic mystical states.

Sargant equates *ekstasis*, "standing outside oneself" with *enthousiasmos*, the "possession by a god."[51] Both follow the same pattern—which is the pattern found in abreactive techniques used during World War II for soldiers with nervous breakdowns, who were preoccupied or "possessed" by their traumatic war experiences. Battle neuroses were helped or cured by provoking intense anger or fear, leading to emotional explosion and collapse. However, such catharsis was not a response to specific traumatic real events (as we might see in Freudian theory), but rather "the creation of intense emotion about almost anything."[52] The pattern of intense emotion, and collapse, followed by a state of peace, is cross-cultural for Sargant. It is based on both Pavlov and Freud and the idea of a common physiology creating common emotional states. For Sargant, the state of ecstasy is basically pathological, but it is one that can have some medical uses. While it causes the person to lose his or her free will in viewing propaganda or while in a hypnotic trance, it may also be therapeutic as a means of getting rid of trauma through a more general form of catharsis.

Cross-cultural ideas of possession are also explored by Richard Castillo, in his article "Spirit Possession in South Asia, Dissociation or Hysteria?" In his view, spirit possession in South Asia is similar to multiple personality disorder in North America; both are spontaneous trance reactions to traumas, especially child abuse. He calls them "parallel trance-related disorders... manifestations of the same underlying pathogenic process" which is understood in culturally specific categories.[53]

He notes that possession by a deity is considered acceptable in South Asia, while possession by another person (such as an ancestor) is considered there as mental illness, and treatment will be sought. He described dissociation theory, understanding trance as a survival behavior in response to life-threatening danger or repeated stressful situations. Such repeated trances can generate separate streams of consciousness and eventually "separate conscious entities with their own sense of identity, behavior, memories and desires."[54]

While "possession disorder" is an indigenous cultural category, he notes that it can be compared to the idea of multiple personalities in Western psychology. American multiple personality disorder or dissociative identity disorder patients have been described as having at least one personality which is protective.[55] South Asian possession also has such figures, in the form of benevolent supernatural beings, which can protect the personality from malevolent supernatural beings. Castillo notes that in North America, psychoanalysts assume an originally unified self which has become fragmented; thus, treatment is aimed at the reintegration of the personalities. However, in India, traditional healers assume that the patient's body has been invaded by external spirits, and the goal is not integration but rather ridding the patient of the invading spirits by means of exorcism. Thus, though a similar pathology is in play, both the healing process and the goal are expressed in different ways.[56]

From the Western analytic perspective, all forms of possession are equally expressions or idioms of distress and are usually interpreted as responses of trauma. Even God-possession, which is highly valued in India and considered to be a desirable religious state, has generally been interpreted as a symptom of dissociative disorder. For some psychoanalytic anthropologists, the debate is over whether there are more and less useful forms of dissociation, or pathological and non-pathological forms.

One well-known anthropological debate over the nature of possession trance was between Melford Spiro and his student, Gananath Obeyesekere. In his book *Gender Ideology and Psychological Reality*, Spiro describes his theory of cultural reproduction and internalization. It includes biological, social, and psychodynamic forces. Cultural internalization describes how acquired knowledge becomes meaningful: cognitively, emotionally, and motivationally. Individuals are active participants in such inculturation, rather than passive instruments. They practice selective filtering or "psychological preadaptation" based on biology and experience, and this sort of internalizing allows people to partially fulfill unconscious wishes and desires. For Spiro, ecstatic experience is a form of psychopathology, in which people confuse internal reality and mental objects, fusing cultural beliefs with direct experience. Religious beliefs are acceptable, but when they are experienced, they become pathology.[57] Ecstasy is impaired reality testing, during which people cannot distinguish between ideas and physical events. Spiro assumes that a distinction between symbol and referent is the norm and indeed the only acceptable state. This is why one may speak to a God, but not actually see him or have him speak back. No matter what the culture, if you have an ecstatic experience of a God, it is a symptom of psychopathology.

Obeyesekere, on the other hand, emphasizes the flexibility of public and private symbols. In his book *Medusa's Hair*, he explores the lives of ecstatics in Sri Lanka, whose interpretations of their lives in mythic and religious terms are deemed valid by the local community. He discusses "cultural symbols whose primary significance and meaning lie in the personal life and experience of the individual."[58] Obeyesekere is interested in the ways that symbols may be used to transcend psychosis, for a symbol can tame a psychic threat, bringing it under the control of ego and culture. Ecstatic experiences can be useful when they objectify and externalize threats, generating a "symbolic idiom that links the individual to both society and culture."[59] It is the "work of culture" to create such symbols, and when they operate at both the personal and cultural levels, their function can be healing rather than destructive. An individual's conflicts are expressed and integrated into symbolic form, and such experiences can justify religious innovation. Obeyesekere notes that "new knowledge has been invented and new cultural rules formulated through ecstatic techniques."[60] Thus, ecstatic experience can also heal psychosis. In some cases, it can help to integrate the ecstatic into the culture, and help to deal with conflicts, by turning an individual symptom into a culturally valued symbol.

However, Spiro critiqued this perspective, and he believed that the ascetics studied by Obeyesekere were simply confusing internal and external realities. This confusion is the basis of the "impaired reality testing" that is the evidence of psychopathology for Spiro. It comes from the common psychoanalytic view that the Western construction of mental reality is the only legitimate one, with mind and external reality clearly differentiated. This view is criticized by Jason Throop in his article "On Crafting a Cultural Mind: A Comparative Assessment of Some Recent Theories of 'Internalization' in Psychological Anthropology":

> In my opinion, the main shortcoming of Spiro's perspective, however, lies not in claiming that there are transculturally recognizable dimensions to psycho-pathology, but rather, that we can speak of something like 'failed reality testing' without first carefully examining how 'reality' is construed within a given cultural system. Indeed, in cultures in which there are long-standing traditions of meditation and ritual that *intentionally* and *systematically* seek to evoke alterations in states of consciousness that often blur the boundaries between inner and outer reality, it seems that turning back to examine the extent to which such varieties of experience evidence 'pathological sequelae' is perhaps a better means by which to judge their pathological or non-pathological nature.[61]

Throop has thus initiated a broader argument concerning psychoanalytic universalism, whether there is a consistent psychological "reality" across all cultures that is being tested and whether good results can arise from debated perspectives.

Along with possession trance, another major area of anthropological research on ecstasy has been shamanism. There has been debate on whether the shamanic trance is a pathological state and indeed whether one can speak of a cross-cultural category of shamanism.[62] Perhaps the best-known writer on the pathological interpretation of shamanism is George Devereux. In his article "Normal and Abnormal," Devereux argues that there are two forms of unconscious, the ethnic unconscious and the repressed or idiosyncratic unconscious. As he states of the ethnic unconscious,

> The "ethnic unconscious" is that portion of the total unconscious segment of the individual's psyche that he shares with most members of his given cultural community. It is composed of material that each generation teaches the next one to repress, in accordance with the basic demand patterns of the prevailing culture. It changes as the culture itself changes... The material composing the ethnic unconscious is maintained in a state of repression by means of various defense mechanisms, usually *strengthened and often even provided by cultural pressures.*[63]

However, when the person's defense mechanisms are inadequate, due to childhood traumas or other stresses, cultures may create an "official, though marginal" recognition and sanction for claimed supernatural and ecstatic states. While these may be accepted by the culture, Devereux notes that "the institutionalization- or excessive elaboration- of an abnormal impulse emanating from the unconscious portion of the ethnic character must not blind one to its culturally dystonic character."[64]

Devereux interprets shamanism as a response to stress, both idiosyncratic and culturally typical, and this stress response is found most clearly in shamanic illness. The responses to stress can evoke cultural defenses (such as ecstatic visions, revelations, or initiation rites with palliative "autotherapeutic" qualities) or ethnic neuroses or psychoses. These are defenses against difficult personal and social situations, in which the shaman claims a "sacred" disorder. Devereux defines the shaman as "a person whose dominant conflicts lie in the realm of the unconscious portion of the ethnic personality," and those who refuse shamanic calls are known to go mad or commit suicide.[65] Thus, Devereux's opinion of the shaman is a negative one: "Briefly stated, my position is that the shaman is mentally deranged." However, he is deranged in a particular, culturally sanctioned way. As Devereux notes:

> The crucial difference between the shaman and the "private" but recognized hysteric or psychotic lies in the fact that the *shaman's conflicts are characteristically located in the unconscious segment of his ethnic personality rather than in the idiosyncratic portion of this unconscious.* Unlike the "private" neurotic or psychotic, he does not have to evolve most of his symptoms spontaneously.... The shaman is psychologically ill for conventional reasons and in a conventional way... Shamanistic treatment tends, moreover, to reproduce both the morbid process itself and the subjective process of cure. This fact explains why the shaman's patient experiences the shaman's actions as reassuring.[66]

According to Devereux, the shaman does not actually cure the patients who come to him, but rather provides a set of culturally recognized defenses, a different set of symptoms, "substituting a taboo for the tic." As in Obeyesekere's approach, idiosyncratic symptoms are changed to cultural symptoms. Shamanic trance is the process of repatterning traumas in culturally acceptable ways. When the ecstatic visions are brought into the culture or acted out (as in the case of Black Elk, who publically enacted his dream vision), this is really "the typical vicious circle of psychopathology," involving both primary and secondary defenses. Devereux noted how shamans are "deputy lunatics" who hold the madness of the tribe, allowing others to maintain a semblance of psychological stability.[67]

In his portrayal of shamans as isolated and marginally sane people, who were tolerated rather than valued by the culture, he differed from other writers of shamanism who understood shamans as central to the culture. He was perhaps most famous for his disagreement with the historian of religions Mircea Eliade, whose book *Shamanism: Archaic Techniques of Ecstasy* portrayed shamans in a more positive light. Eliade understood shamanic illness as a temporary sign of election to a shamanic role. In his chapter on shamanism and psychopathology, Eliade notes a range of anthropological interpretations of shamanic illness, including neuropathy, neurasthenia, epilepsy, and hysteria. However, he concludes that the shaman is "a sick man who has been cured, who has succeeded in curing himself."[68] The shaman is a master of ecstasy, who learns how to control such states for the benefit of his community.

While Eliade is arguably the most famous historian of religion of the twentieth century, the field of History of Religions has moved in a different direction after his death. It has become more skeptical of ecstatic states, pathologizing them in a variety of ways.

History of Religions/Religious Studies

The fields of Religious Studies and History of Religions have grown up in tension with their parent field, Theology. While Theology has emphasized faith, Religious Studies has emphasized criticism. Both, however, have been skeptical about ecstatic states.

Ecstasy has been condemned for supporting illegitimate and oppressive social systems. Grace Jantzen's *Power, Gender and Christian Mysticism* critiques the social construction of mysticism, arguing that it is bound up with ideas of authority and gender. Her writings on ecstatic experiences among medieval female saints emphasize the institutions of religious power and the long history of struggle between competing power blocks vying for religious domination. The experience of women mystics was a potential source of power, which was understood by Catholic theologians as threatening to the male hierarchy of the church and a subversive force which had to be controlled. Otherwise mystics, like women and other oppressed groups, would resist the power and authority of the church. Many female ecstatics accepted this situation and went along with the church rather than responding to the oppression.

Jantzen criticizes writers on mysticism who focus on experience, as she feels that they should focus instead on politics, not dividing the personal and political. Unless there is political protest, people adapt to bad situations rather than challenging them. The "privatization of spirituality" reinforces the societal status quo.[69] However, she does note that some women mystics were able to use their ecstatic states as a basis for religious authority. Jantzen claims that women had ecstatic visions because they "had internalized the low esteem in which they were held," and lacked formal education and ecclesiastical position.[70] Female mystics were feared by men for their sexuality, and they were forced to seek control over their lives by fasting and religious practice. But female holiness was bought at the price of accepting gender stereotypes, and identifying with food, flesh, and suffering, a "price no man would ever be expected to pay."[71] We might note that this is, of course, except for Jesus, who was identified with all three.

Heretics were mystics who had failed, based on the judgment of religious authorities who monopolized the decision as to whether an ecstatic experience was of divine or demonic origin. Jantzen notes that women were called heretical when they asserted authority and would not be submissive and humble toward the church; this could result in their ideas being suppressed, or in their being killed.[72] Jantzen claims that there is pathology in the ecstatic experience itself, which has been warped by the dominant religion, and in the social order surrounding it, in which the church had total control over nuns and female mystics. The problem for the mystics and ecstatics was accepting the institutional Church and not resisting it enough and being willing to follow the power structure. We might note that it is difficult to know very much about female visionaries who did not go along with the church, as we primarily have the records that the church and its hagiographers collected.

Jantzen accepts the positions of Marxism and feminism, in her concern with oppression and social justice. From these perspectives, ecstasy is about social authority rather than inner experience. She states clearly that there is no such thing as a mystical "essence."[73] Ecstasy exists only as a symbolic way of manipulating power, and its truth claims are false ones; ecstatic visions are fantasies through which women gain status in a male hierarchical world.

Ecstasy has been interpreted as a tendency toward self-destruction in Ariel Glucklich's book *Dying for Heaven: Holy Pleasure and Suicide Bombers*. Glucklich states that the first major warning sign of religious war and nuclear attack is whether a religious leader has mystical or ecstatic experiences.[74] Secondarily, one looks for discussion of love. As he notes, "A group that is both ecstatic and reformist is likely to threaten the foundation of the host state."[75] Religious ecstasy is "annihilative," inclined toward destruction of the mundane world because of its experiences of God or heaven. Thus, because they speak most of ecstasy and love, he believes that it is most likely that Sufis would pose the greatest danger of nuclear war in the Middle East.[76]

Martyrs and suicide bombers may be ecstatics in themselves, or their theatrical aspects of martyrdom may make their supporters feel uplifted, awestruck, ecstatic, and joyful. However, Glucklich claims that ecstatics and mystics who think that their experiences reflect a transcendent state are wrong. They are subject to a "hedonic attribution error," believing that their joy is from divine love instead of a "functional and chemical brain state." This is described as the "Prozac effect" and results in states of joy, bliss, rapture, and love. It feels transcendent, but it is chemically induced, and a product of natural and cultural evolution. Religion is an "alchemical hedonic device," transforming pleasure into happiness.[77] Such pleasure is based on ignorance, as Glucklich notes, "the actions of martyrs, mystics and suicide bombers are both profoundly hedonic and misperceived."[78] The concept of "annihilative ecstasy" links annihilation of the individual soul in God's love with annihilation of countries by means of weapons of mass destruction. Ecstasy thus becomes a precursor to physical violence, a pathological state which leads to psychological and physical destruction, homicide and suicide.

Most religionists are not trained as psychoanalysts, yet psychoanalytic categories are often used in Religious Studies. Most psychoanalysts would say that training is necessary, to avoid problems like projection and countertransference. Yet approaches like psychobiography and case histories with "retrospective diagnosis" have become popular techniques of pathologizing ecstatic states in the field of Religious Studies. While there have been a range of books linking ecstasy with trauma, here we shall include as examples Hollywood on ecstasy as fetishism, Bell on ecstasy as anorexia, and Kripal on ecstasy as repressed homosexuality.

In her book *Sensible Ecstasy: Mysticism, Sexual Difference and the Demands of History*, Amy Hollywood deals with ecstasy as fetishization and body symbolism. She notes that ecstasy was interpreted as delirium and other pathologies by the early psychologists Janet and Charcot, and she mentions the problem of pathologizing gender and ecstasy:

"Usually the term 'pathological' refers to the forms of affective, erotic and body mysticism associated, in the medieval and modern periods, with women."[79] She points out that visionary and emotional mysticism is often associated by modern scholars with female mystics, while the masculine style of mysticism is more intellectual, speculative, and often explicitly anti-visionary.[80] Women were oppressed by the church in the Middle Ages and were assumed to be more open to possession, visions, and auditions and more focused on bodies rather than texts. Their ecstatic states were condemned as escapism, by a church that emphasized action over mere contemplations.

Hollywood examines French thinkers who were strongly influenced by Freudian psychology and their views of mystical and ecstatic experience. For Georges Bataille, Jacques Lacan, Simone de Beauvoir, and Luce Irigaray, mysticism is "an encounter with human suffering, illness, death and mortality, which is in itself an encounter with the sacred or divine."[81] For these writers, the sacred is non-theistic. The bodies of mystics, especially female ones, are the sites for projections of disease, pain and death, as well as repressed desires and traumatic emotions. On the whole, Hollywood finds ecstatic experience to be a reaction to the loss of traditional ritual: "The modern fascination with the mystical exhibits a nostalgia for a time when there were ritual means to deal with the traumatic effects of loss, limitation and death."[82]

The body of the ecstatic person is a sort of screen, on which both personal and social tensions are projected. This may be seen in the ecstatic imitation of Christ's suffering, in ascetic practice, in ecstatic anguish which seems more like misery than the presence of God. Indeed, for Bataille such states are beyond God, they are inner experiences which transcend the notion of a personal God, and are reached through immersion in pain. For Hollywood, the "hysterical and/or fetishized body" of the ecstatic holy woman is a site for tensions: clergy vs laity, orthodoxy vs heresy.[83] Such states are further pathologized with the addition of erotic imagery, which was usually interpreted as a hysterical symptom. Fetishism was originally an anthropological category found in colonial history and later a Freudian symptom of castration anxiety. But for Hollywood, the term is used more broadly, as a displacement of a range of psychosexual and social anxieties. Rather than originating from a single trauma, it has multiple sources in historical issues of power and oppression. As she states, "We need both to historicize and politicize our reading of fetishism."[84]

Interpretation is important, for in medieval times the church would often interpret the ecstatic experiences of the female mystic in a different way than she did. Male clergy and hagiographers were believed to be capable of understanding the female mystic's experiences more accurately, as they were more educated. Her claim could be due to divine or diabolical possession, or to disease (especially imbalances of the humors, though womb problems were also considered to be possible). Because a women's access to religious authority came primarily through her ecstatic experiences, interpreters and hagiographers wanted control over the spread of her ideas. As in Jantzen's concerns, her mystical experiences were interpreted as supporting the church hierarchy: "The hysterical and fetishized mystical saint is a divinely possessed, ecstatic figure who provides a powerful and visible safeguard of male priestly authority."[85]

In Hollywood's writing, the pathology is partly social, but it really appears in the suffering body of the ecstatic person, as a reflection of both her own and society's sufferings. Though the mystics spoke of the joys of Christ's love and visions of salvation, the focus in Hollywood's book is on their bodies rather than their statements of joy and love. According to her interpretation, ecstasy is not joy, but rather sorrow and bodily suffering, a mass of human conflicts and projections rather than the vision of God that the ecstatics themselves describe. As Hollywood noted in terms of the medieval hagiographers, the interpretation can easily become more important than the original experience.

In his book *Holy Anorexia*, Rudolph Bell is also interested in the psychological and biological problems that come with ecstatic states. He too examines medieval female mystics, and notes that women were more subject to "inner" problems, for in women evil was internalized, while for men it tended to be an external problem, associated with sin and impurity.[86] His concept of "holy anorexia" on the part of mystics and saints involved the suppression of physical urges and basic feelings (fatigue, sexual drive, hunger, pain) which "frees the body to achieve heroic feats and the soul to communicate with God."[87] The female mystics who fasted and had ecstatic experiences were really anorexics with unbearable conflicts over dependency, for whom every day was a desperate choice, a daily, relentless, psychically torturing challenge for a girl who feels hopelessly inadequate and ineffective.[88] Such female mystics would seek to be beautiful in God's eyes rather than man's, and anorexia allows them to be spiritual champions and brides of Christ. Insecurity turns to "fanatical, compulsive devotion," following the wider social pattern of heroic, ascetic masochism.[89] While such masochism can be found in both genders, Bell focuses on the female situation.

He discusses the pattern of the religious girl who runs away to a convent to avoid marriage, is tormented and unable to eat, but later "recovers" and becomes active.[90] There are initially more negative visions, full of rage, "narrowly escaping schizophrenic depths," and later more positive ones.[91] He focuses upon the thirteenth century, and the ways that saints chose to imitate Jesus in the wilderness, fasting forty days and nights as controlled penance, and otherwise trying to live an ascetic life. Such choices might be motivated by childhood trauma (especially the death of a loved one), hatred of sexuality, or following a social ideal to be emulated by pious women. Thus, for Bell, the ascetic lifestyle and ecstatic visions of the female mystics were symptoms of disease, and a religious life was only a superficial veneer over problems of insecurity, depression, obsessive–compulsive behavior, masochism, and schizophrenia. Ecstasy came from crises over autonomy and conflicts over sin and was a culturally acceptable way of dealing with traumatic experiences. Fasting and anorexia were ways of expressing conflicts, and their ecstasies were fantasies in which conflicts were resolved.

Pathologizing in the context of Eastern religion is found in Jeffrey Kripal's *Kali's Child: The Mystical and the Erotic in the Life and Teachings of Ramakrishna*. His characterization of the Hindu saint Ramakrishna Paramahamsa in this book evoked a strong protest from the Indian media and community and became part of an ongoing controversy over the nature of religious ecstasy in India. Ramakrishna was a nineteenth-century ecstatic and priest who ritually worshipped the goddess Kali as the Divine Mother. He is also understood by his modern followers to have been a Vedantin, one who accepted the idea that many paths to God were legitimate, and that all religions were symbol systems which led to the same divine goal. Kripal portrayed Ramakrishna not as a divine sage and enlightened person, nor as a God incarnate (as he is often understood to be in India, with the title Ramakrishnadeva), but rather as a child who had been sexually traumatized and a lifelong frustrated homosexual who was abused by his relatives and gurus. The Indian protest focused upon the fact that these claims were not based on events described in Ramakrishna's biographies, but were rather deduced by Kripal from what Ramakrishna's biographies did not say. The controversy involved the question of how much may legitimately be "read into" texts and when implication turns into projection.[92]

Kripal described his book as finding a "fusion of mysticism and eroticism" in Ramakrishna's life. The saint's mystical trances were described as ego-defenses against past traumas, so that true knowledge was repressed sexual desire and abuse, while mystical states were a sort of ignorance, which hid the truth. The scholar is thus wiser than the saint, knowing that sexual trauma is a deeper truth than religious Enlightenment. The closest that we see to a fusion of these ideas is the concept that both ecstatic trance and repressed sexual impulses are unconscious, so they are in that sense equivalent.

Kripal claimed that Ramakrishna's experiences were "profoundly, provocatively, scandalously" erotic,[93] and that Ramakrishna was a "conflicted, unwilling homoerotic tantrika" who was full of shame, disgust, and fear. The saint's ecstatic trances were only defense mechanisms, ways to avoid remembering his past traumas and current frustrated desires for his disciples. As Kripal phrases it, Ramakrishna's ecstasy "looks as much like a defensive trance as a mystical state."[94] His beliefs about Ramakrishna's repressed homosexuality and his literal interpretation of Ramakrishna's joke about "defecating ecstasy" were very controversial.[95]

Citing the psychoanalyst Jeffrey Masson, Kripal states that there are three major traumas: sexual seduction, overt or covert aggression, and the loss of a loved one.[96] Kripal finds all of these in Ramakrishna's life; however, they are hidden in ecstatic states rather than described by biographers. For instance, when biographies state that Ramakrishna spent extended time learning meditation practices from his guru Totapuri in a meditation hut, Kripal stated that this meant that his guru had been sexually attacking him.[97] Ramakrishna's success in gaining the state of *samadhi* or insight was not a sign of attainment, but rather an escape from a sexually abusive guru. The lack of descriptions of sexual and violent events in the biographies of Ramakrishna is read by Kripal as paradoxical proof of their existence; the fact that they are not mentioned in any biographies proves that Ramakrishna is trying to hide them.

For Kripal, Ramakrishna's ecstatic states represent symbolic castration and homosexual repression, and the goal of such states is the resolution of psychosexual crises.[98] Even Ramakrishna's blood dysentery (a disease widespread in rural West Bengal) is interpreted as a psychosomatic statement. His "secret talk" (*guhya katha*) or spiritual guidance for disciples was not meditative advice, but rather ways of resolving sexual conflicts, and struggling with the "horrors of his past."[99] He was celibate not because he chose this as a spiritual path (and it is traditionally a respected choice in religious people in India), but because he was unable to have sex with women, and secretly feared them. For Kripal, it is a dogmatic fact that religious ecstasy reenacts traumatic events, allowing the person to escape from memory by entering eroticized trance states. Such claims can be made by the "psychoanalytically trained observer," because he or she understands symbols and rituals better than the performer does.[100] Like Hollywood, Kripal constructs ecstasy as traumatic in origin and focused upon the body.

Kali's Child is a psychobiography, a genre of biography that evolved from case studies by practicing psychoanalysts. But psychobiography has migrated out of analytic practice and is now used as a form of literary theory. It is often used so carelessly in academic writing that the approach has become popularly known as "the maligning of exemplary figures," in which all virtues are side effects of vices, and major accomplishments are understood as reactions to childhood experiences and the results of intrapsychic dynamics. In some cases, the genre can become a form of academic "yellow journalism," focusing on imagined sex and violence in the person's life. This can be shown through "retrospective diagnosis" across cultural and religious ideologies, when Western suspicions reach over time and space, diagnosing psychotic disorders in the religious experiences of people in other cultures.

As we can see, many historians of religion have been influenced by psychoanalytic theory. We also have trained psychoanalysts who have written about the topic, having examined people in ecstatic states and theorized about them.

Psychoanalysis and Ecstasy

Psychoanalysis has long been suspicious of ecstatic states. It has condemned ecstatic experience as narcissism, as an escape from responsibility, and as a regressive behavior returning people to infantile helplessness. Psychoanalysts writing on ecstatic religious figures have interpreted ecstatic states as defense mechanisms against childhood traumas, narcissistic grandiosity, and attempts to reunite with the mother. Visions of deities show repressed sexuality and aggression and allow formerly hidden desires for power to emerge. For Freud and many of his followers, the oceanic feeling of union is a form of regression, while the vision of God represents a projected earthly father. Ecstatic states may be interpreted as dissociation, depersonalization, derealization, and symptoms of post-traumatic stress disorder. Early psychoanalytic studies on ecstasy followed Freud closely and were perhaps even less sympathetic. Ecstasy has been variously described as equivalent to or motivated by hysteria, psychotic states, artificial catatonia, dementia, manic denial, homosexual urges, and incest fantasies.

The pathologizing of ecstasy in psychology has had many contributing writers, but it is traditional to begin with Sigmund Freud. He laid the groundwork in his 1927 book *The Future of an Illusion*, where all religious claims were dismissed as illusory wish-fulfillments, a way of getting the masses to agree to work and suppress their instincts (which he lists as desires for incest, cannibalism, and murder). External coercion becomes internalized, people develop superegos, and the religion supports the culture. Religion protects mankind from helplessness, makes finite life tolerable, and gives people faith because they are forbidden from questioning the sacred texts of their ancestors. As the "universal obsessional neurosis of humanity," religion can bring hallucinatory bliss, but not truth. Its claims unite people, and this is useful for cultures, for in Freud's view "every individual is virtually an enemy of civilization."[101]

This tension between the individual with his or her instinctual drives, and the culture which forces moral behavior and thus necessitates neurosis, is further developed in Freud's *Civilization and its Discontents.* In this book, Freud discusses the ecstasy as the "oceanic feeling," interpreting it as an instance of primary narcissistic union between the mother and the infant. When adults have this feeling, they are regressing back to infancy and seeking the bliss that infants feel. Metaphysical and transcendent claims about such a state are false and come about through cultural elaborations. Mystical and ecstatic states are infantilism and regression, pathologies involving denial of the real physical world in favor of the quiet and security of the intrauterine state. As Freud describes it, the oceanic feeling is

> A sensation of eternity, 'a feeling as of something limitless, unbounded- as it were, oceanic.' This feeling… is the source of the religious energy which is seized upon by the various churches and religious systems, directed by them into particular channels, and doubtless also exhausted by them.[102]

Freud's interest in the oceanic state came about through letters and conversations with the writer Romain Rolland, who coined the term "oceanic feeling" from doing research on the Hindu saint Ramakrishna

Paramahamsa.[103] Rolland spoke of the sense of "totality of being," how one's essence was Being Itself at the moment of ecstasy.[104] Freud did not agree with these claims and felt that at best such a feeling was a "benign vestige from early phase of pre-Oedipal life."[105] Freud also interpreted ecstasy as part of the death drive, which was opposed by Eros or libido. A value on inner experience showed an attraction toward death.[106]

Later Freudians also came to be interested in the connection between Hindu religious states of consciousness and Freud's "oceanic feeling." The Sanskritist and psychoanalyst Jeffrey Moussaieff Masson understood ecstatic states in the Indian tradition to be rooted in childhood trauma. He stated that ancient Indians could not acknowledge emotion, and this "massive denial" was "one of the fundamental disguised themes of almost all ancient Indian writing."[107] Practices like yogic meditation and renunciation were struggles against trauma and adolescent sexuality, but no matter what yogis and monks claimed, they were scarred from "this form of deep emotional surgery." Masson claimed that renunciants who seek Enlightenment have an obsessive concern with depersonalization, an internalization of aggression which is a sort of psychic suicide.[108] He also argued that experiences of Ultimate Reality are really ego-defenses using a denial of reality: Since nothing physical exists during such states, therefore there is no pain. Claims of memory of previous lives are denials of one's own earlier life, one's childhood. Ecstatic travel to other worlds is a fantasy of annihilation, and seeking the ocean of *Brahman* is a desire for death: "the longing for the ocean as an overt symbol of eternal peace is always a covert reference to the violence of suicide."[109]

Masson identified specific Hindu saints whose ecstatic experiences were responses to trauma. He focused on Ramakrishna Paramahamsa, as did Kripal later on. He too stated that Ramakrishna suffered sexual trauma in childhood, though there was no clear evidence of this. This was obvious to him, as he assumed it to be the origin of all ecstatic and mystical experiences, shown in "the recovery of (or more accurately, with the inability to recover) early affect-laden memories... almost inevitably of traumatic events. In these contemplative states, in place of the memory (whose reliving would be painful) the mystic substitutes a blissful, denying experience."[110] In his article "The Psychology of the Ascetic," Masson notes that the scholar is the opposite of the ecstatic in his control over repression:

> I believe that the scholar, the erudite mystic is in greater control of his repressed wishes- his repression is more tenacious; whereas in the ecstatic,

> the repressed has returned, the barriers have broken down, and he is closer to what analysts call primary-process thinking. Hence, for a study of the underlying motives that impel a man towards mysticism in general, it is this form that allows us a clearer, more sustained view into the unconscious. The studiousness and sobriety of the scholar are defenses erected against the impulse to behave in exactly the way the ecstatic does.[111]

While scholars in the USA are perhaps ascetics metaphorically, in India becoming an ascetic or renunciant was quite literal and traditionally a valuable thing to do. From Masson's standpoint, however, there is no positive aspect to such a choice. Going out to seek Enlightenment is "like the little child who packs his bags and leaves for the open country, all the while desperately awaiting the sign from his parents that will allow him to abandon his spite and rush to the waiting arms of a loving parent."[112] Masson states that *all* ascetics suffered massive traumas in their childhoods, and that all addicts have suffered such losses as well.[113] The seeking of ecstasy in renunciation is simply an attempt to escape such traumas. Masson notes that, in his opinion, the "finest study ever made of mystical states of consciousness" was Bernard Lewin's *The Psychoanalysis of Elation*. Thus, we turn to Lewin.

In *The Psychoanalysis of Elation*, Lewin discusses major theories of ecstatic states from the first half of the twentieth century. The range of human experience that has been equated with pathology is quite striking; for Lewin, nothing is what it seems.

Lewin defines elation as "joy, bliss and euphoria," but he interprets these as mania.[114] Mania (a term which goes back to the ancient Greeks) is used "for a variety of excitements, delusional states and other conditions."[115] Lewin's emphasis is on the manic attack or hypomania—the undelusional, overactive and elated state. He finds it to be usually associated with depressive states, a "circular psychosis," with most later psychoanalysts focusing on the depressive side. For Lewin, hypomania is an obsessional neurosis in which the patient is cheerful, optimistic, and witty, but his or her disease is shown by resistance to the analyst's observations.[116] Cheerful patients were avoiding what the analyst considered to be underlying problems. Lewin found such states to be precipitated by significant life events (in woman, especially sex and childbirth), but elation was an "unstable defense." It hides the basic human impulses, which are "cannibal impulses" and the hunger and thirst for new experiences which represent new "food for thought."[117] These originate in "psychological greediness," the desire for new ideas and experiences which is a sign of pathology.

These states of joy and interest in new ideas that Lewin observed are really "technical elation," which deny or intellectually repudiate the patient's underlying problems. Elation is a "screen affect," similar to a "screen memory," where the person only thinks he or she is happy, but such happiness masks deeper anxiety. These anxieties are revealed by the person's "elated denials," which show that the patient's narcissism has been offended.[118] Such anxieties are assumed to go back to infancy.

Lewin states that the basic desires of the infant are to eat, to be eaten, and to go to sleep.[119] Sleep and death are both psychologically equivalent to being devoured. Acceptance of such desires results in ecstasy, while the denial of such desires leads to mania, or "hypomania proper."[120] He defines ecstasy as "a repetitive fulfillment of the oral triad wishes to relax and to be devoured into happy sleep."[121] Brief ecstatic nights may precede the overactive and happy manic state; the person may have intense dreams, visions, and delirious experiences. But the manic state is the conflict between the ego's desire to sleep and stay awake, to die and to live. What appears to be happiness is really conflict, in particular that involved in "active cannibal orality."[122] The manic person thinks that he or she wants happiness, but this is just a desire to return to early infancy. Bliss is like a "blank, erotic, orally satisfying dream" which repeats the infant's feeling of union at the mother's breast.[123] It reflects the reality of the "breast experience" and the "primitive narcissistic trust" in sensory experience.

Ecstasy is thus sensory for Lewin, and repeats a past memory, even if it appears not to resemble the memory. Ecstasy may also involve an "immortal superego figure" and a regressive sense of immortality which "repeats the infant's ignorance of the passage of time."[124] Indeed, the blank state of satisfaction that follows nursing is the true origin of such religious goals as heaven, the oceanic feeling, immortality, and nirvana.[125] The vision of God or heaven is also the psychological equivalent of suicide, for it only hides the negative aspect of death or annihilation.

Thus, the desire for peace, joy, and security is regression. However, activity in the world is also a problem. The neurotic hypomanics who wish for fame, wealth, and success in the world really want immortality and identification with their dead parents.[126] Though active people's "social efficacy is beyond dispute," that is irrelevant, for they are only defending themselves against the fear of death. Wealth is the expression of an unconscious fear of starvation and being eaten. In Lewin's memorable phrasing, "success avoids the danger of eating oneself up and of living on one's own fat."[127]

Elation, the happiness reflected by both contemplation and action, is a false emotion, as delusion is a false idea, and hallucination is a false perception. Though elation has an emotional sense of reality, it is not real. A psychoanalyst cannot trust the reporting of such emotion, for it only hides the fear of death and anxiety. One may tell the manic person by the hunger for new ideas and impressions, and by the desire to learn and grow, which is actually regressive. As Lewin states, "Hunger and the acquisition of knowledge coincide at the breast and have the same object."[128]

As we can see from the psychoanalytic perspective, it is difficult to find any human activity that might not have a pathological basis. Ecstatic states were a particularly clear target. By the 1970s, we have the generalized notion of schizophrenia (which at that time referred loosely to all psychotic illness) which stemmed from childhood experiences, with a few "anti-psychiatry crusaders" like R. D. Laing and John Perry, who claimed that psychotics could be on mystical voyages, and that ecstasy and psychosis were not necessarily the same.[129] However, they had little success in battling against the dominant medical model of ecstasy as disease.

But the 1970s also had a shift from psychiatry to biology, with critiques of Freud being widely published. Freud was portrayed in the popular press as a bumbling Inspector Clouseau, an incompetent, full of theoretical absurdities. It was an ongoing battle; in 1976, the Group for the Advancement of Psychiatry supported Freud and described religious experience as "borderline psychosis," and "a regression, an escape, a projection upon the world of a primitive infantile state."[130] Though the strict Freudian perspective of earlier psychoanalysis has declined, the use of therapeutic drugs has increased, and we see more modern pathologizing theories from the medical perspective, an approach sometimes called biomedical reductionism.

Biology, Psychiatry, and Medicine

From the medical perspective, ecstasy is caused by physiological and biochemical problems, ranging from nutritional deficiencies to diseases of the brain and nervous system anomalies. As an example of the dietary theory, we have the argument that ecstasy is a result of calcium deficiency, which is especially marked in possession cults.[131] This idea originally came from Anthony Wallace's analysis of Arctic hysteria and was taken up by Alice Kehoe and Dody Giletti, who extended it more generally to women in possession cults. They argued that convulsions and cognitive disorganization that came from malnourishment would incline people, especially women, toward trance states. They write:

> Our suggested explanation for women's predominance in spirit-possession groups derives from demonstrable relationships between diet and behavior. There is a strong correlation between populations subsisting upon diets poor in calcium, magnesium, niacin, tryptophan, thiamine, and vitamin D, and those practicing spirit possession.[132]

This theory has been applied to possession groups worldwide, and social scientists have examined the relationship of stress to diet, as well as the role of dietary restriction, in creating the "indifference to harm" sometimes found in trance states. Because the theory does not take into account religious and cultural factors, this argument has been criticized by anthropologists as biological reductionism.

A very different argument for the biological origins of ecstatic experiences came from Julian Jaynes, the Princeton psychologist, who argued that ecstasy originated from the split in the bicameral or two-chambered mind. In his 1982 book, *The Origin of Religion in the Breakdown of the Bi-Cameral Mind*, Jaynes describes early mankind as having a brain split into two separate spheres. The "god-side" of the brain, in the right hemisphere, spoke to ecstatics in the form of auditory hallucinations. In mankind's early years, the parts of the brain were not connected as they are today and communicated over a gap. Jaynes associates communication between the cranial hemispheres with claims by ecstatics that they are speaking to the dead, communing with statues inhabited by gods, and receiving messages in temples which housed gods and oracles.

Before the year 1000 CE, Jaynes states that auditory hallucinations were considered the norm. But they have died out in the modern world as the human brain structure has changed, and the two sides of the brain have become more connected. His theory is that ancient Greece, Babylon, Egypt, and the Mayan and Incan kingdoms were all built by people who were not "conscious" in the modern sense, and were capable of ecstatic communication with gods and spirits. However, vestiges of the older kind of consciousness remain, in such experiences as channeling, glossolalia, possession trance, automatic writing, and hypnosis. This older relationship of the brain hemispheres may also be found in some types of schizophrenia, when there is a reversion to the older forms of brain functioning.[133] This theory is another form of the argument that primitive people are physiologically different from civilized people; in this case, it is their brain structure that is different. Jaynes considers all forms of mystical and ecstatic states to be archaic, mental relics of the past.

Canadian researcher Michael Persinger is probably the most well-known figure to describe ecstasy as a problem of the brain and nervous system. He argues that ecstatic states result from anomalous electromagnetic fluctuations within the brain, from solar flares, seismic activity, and even electrical devices. Whatever the source of the mental disturbance, it triggers "microseizures" which resemble epileptic seizures, and these generate altered states—including religious visions, out of body experiences, and UFO abduction episodes.[134] Persinger has tested the induction of such microseizures on healthy volunteers, stimulating their brains with his "God machine," and he claims anywhere up to a 40% success rate (a rate which has not been replicated by other researchers). A variety of brain disruptions might cause ecstatic and mystical experiences, including head injuries, strokes, epileptic seizures, drug use, and physical traumas. The disruption could induce a "sensed presence," which is interpreted as angels, fairies, ghosts, muses, and gods.[135]

Persinger speculates that the left and right brain hemispheres used to be more poorly integrated and less developed thousands of years ago, when the left hemisphere (which specializes in language and identity) attributed signals from the right hemisphere as coming from external sources. Thus, much like Jaynes, he finds the origins of religion in the gap between brain hemispheres.[136] For Persinger, ecstasy most likely comes from injury to the left hemisphere of the brain, and he speculates that all meditators who have religious experiences could have low-grade epilepsy.[137] Ecstatic states are due entirely to neurological problems, and as such are physiological pathologies.

There have been many debates about the exact biological locale of religious experience, with its origins claimed to be in the neural circuits of the brain's temporal or parietal lobes, the general right hemisphere of the brain, the corpus callosum, and the synaptic cleft between neurons (in the various neurotransmitters).[138] Most medical writers have described religious ecstasy as a state which could be fixed with the right chemical given to the patient, and new antipsychotics are being developed.

From a conservative psychiatric perspective, ecstatic states are both dangerous and undesirable. As one psychiatrist noted in informal conversation, he spent his life trying to help patients who desperately wanted to be normal in order to escape psychological pain. The goals of transpersonal and humanistic psychology, having people attain supernormal or peak experiences, were impossible and gave people false hopes. There was no such thing as a supernormal state of consciousness, there were only normal and subnormal states, and people should be satisfied and grateful if they were normal.[139] This idealization of the norm has worked against research on ecstatic states, for these states have been categorized as forms of mental illness. This categorization has hindered cross-cultural understanding in the fields of both religion and psychiatry and imposed Western categories of illness on states that are considered as desirable and highly valued in other cultures.

In his book *Crazy Like Us: The Globalization of the American Psyche*, Ethan Watters notes that Western concepts of sanity and insanity have been sent around the globe. "Over the past thirty years, we Americans have been industriously exporting our ideas about mental illness."[140] Many modern Western ideas about mental pathologies come into conflict with local ideas, concerning both resolutions of crises and altered states of consciousness. The biomedical models of mental illness and trauma came to dominate local ideas of spirit possession, visions, and other "culture bound syndromes" (we may note that Western ideas are not considered to be "culture-bound"). Local conceptions about mental disease and also ecstatic states are dying out. Watters compares cross-cultural anthropologists to botanists in the rain forest, "desperate to document the diversity while staying only a few steps ahead of the bulldozers."[141] Such a loss of cultural diversity is like the extinction of species, funded by drug companies seeking ever more patients to consume their drugs and be diagnosed through their disease models.

Watters notes that patients try to fit their symptoms to the medical diagnoses of experts, which medical historian Edward Shorter calls "illness negotiation" between doctors and patients.[142] Where local experts might deal with stresses in problems in indigenous ways, Western doctors and psychoanalysts have come in with very different assumptions. As an example, in cultures which emphasize avoiding the memory of past evil events, post-traumatic stress disorder (PTSD) counselors who come from other cultures insist that disaster victims relive their traumas. The counselors think this will be good for them, but in their culture, this brings the evil spirits of disaster back to the victims once again. In Sri Lankan sufferers of PTSD, terror was understood locally to make people vulnerable to "the gaze of the wild," a wild spirit which throws people into the trance state of *inna barkama*. One can only avoid this state by silence or by indirect speech, and describing violence is very dangerous. The evil will come back to possess the person, bringing a "terrified heart."[143] Yet describing the violence and re-experiencing it is exactly what the Western therapists assume is the basis of psychological healing. It is a conflict of models of pathology, which can be harmful, even when people are trying to be helpful. As the medical anthropologist Arthur Kleinman notes,

> Most of the disasters in the world happen outside of the West, yet we come in and we pathologize their reactions… We take their cultural narratives away from them and impose ours. It's a terrible example of dehumanizing people.[144]

Different cultures and religions tend to recognize different categories of mental distress. As Watters notes, in Judeo-Christian cultures there is much delusional guilt, people hear the voice of God condemning them in demanding and critical tones. This would be what one might expect from a Bible which has the voice of God speaking to people and, in the prophetic books, usually condemning them. American schizophrenics often identify with celebrities, world-famous and full of wealth and power, who are valued by the culture, while in Southeast Asia, delusions of grandeur are rare, because the culture values humility. Indeed, the delusional content is often associated with the values of the culture. There is less guilt, though there is more concern about violating purity rules. Watters notes that individual symptoms are pathoplastic, varying from person to person, but researchers seek the pathogenic or root cause of the disorder. They seek to weed out the "incidental content" and get to the "essential form."[145]

Yet the local forms of healing, with their acceptance of spirit voices and possession states, often have better rates of recovery from mental illness. Watters mentions a famous international study on schizophrenia by the World Health Organization, which included over a thousand patients from ten countries and continued over twenty-five years. The disease resulted in greater impairment in the industrialized nations than in the Third World, with schizophrenics in Nigeria, India, and Columbia, for instance, less severely impaired than in the USA, Denmark, and Taiwan. While 40% of people in the industrialized West remained severely impaired after treatment, that diagnosis was appropriate for only 24% of patients in poorer countries. There was no convincing or clear explanation as to why people in developing countries have better outcomes than those living in industrialized nations. Janis Hunter Jenkins and Robert John Barrett noted these as very perplexing findings in the area of epidemiology.[146]

Indeed, more and more forms of behavior have been categorized as mental disease in the USA. The psychologist Richard McNally has recently described the development of disease categories in the various editions of the DSM (the official psychiatric listings of mental illnesses). While there were 14 discrete mental disorders listed in the DSM-II of 1972, the 1980 DSM-III listed 211 mental disorders, and the 1994 DSM-IV listed 341 mental disorders. He questions whether psychiatry has discovered over 300 new forms of mental illness unknown in 1972.[147] He also described a 2005 survey of over 9000 adults interviewed by psychiatrists, which concludes that nearly 50% of Americans have been mentally ill at some point in their lives.[148] He mentions the boundary problem described by critics of psychiatry, "expanding the boundary of mental disorder, inappropriately pathologizing everyday life by classifying normal variations in emotion, behavior and cognition as disorders."[149] He believes in the reality of mental illness, but questions how such diagnoses can best be determined. The current 2013 DSM 5 lists over 300 mental disorders.[150]

It may well be that less industrialized countries have more tolerance for unconventional states of mind than the industrialized West. If we take the case of India, it is likely that the ecstatics and visionaries who were highly valued in their social groups would have been interpreted as having some form of dissociative or manic depressive disorder if they lived in the West. Bhakti devotion in India involves extremes of joys and sorrow, and perceptions of gods and their worlds, as well as possession by deities. These states would not be considered as valuable in the modern West, or even acceptable.[151] Some cultures have the boundaries of acceptable mental states drawn much more tightly than others, and the number of people classified as diseased is likely to increase as Western categories become more accepted.

Watters also noted that cultural acceptance of mental difficulties as originating from spirits rather than from disease has many virtues. It lessens the stigma of past bad behavior and makes it forgivable; the person can return to the group with less shame and blame.[152] Local illnesses link people to the history of their culture, and this makes their problems understandable. However, accepting the Western model of mental illness, the disease model, often means violating traditional rules of modesty and bringing treatment that is cold and dehumanized. It makes people appear threatening because the underlying presumption is that their illness is inherent and long term rather than temporary, and caused by invaders who can be symbolically banished. "Mental health literacy" comes to mean accepting an alien model of mind and spirit, which can harm the people it was intended to help.

It does not help that Western paradigms of normalcy, views of human nature, definitions of personhood, concepts of time and memory, and sources of moral authority change over time.[153] States of mind which were once acceptable are now too extreme for industrialized culture, when too much joy becomes mania, too much sadness becomes depression, too much thoughtfulness becomes introversion, and too much ecstasy becomes schizophrenia or dissociation. This is the dominant medical and psychiatric view of ecstatic states in the modern West. The virtue from the business perspective is that, as we find more pathologies, more drugs can be sold for them.[154] Many people can profit from the narrowing of acceptable ecstatic states and expansion of diseases, for "there is money in madness."

As we can see, many forms of pathologizing and condemnation from a variety of academic disciplines have been presented, and in some fields, these are dominant perspectives. One might assume that Theology would be a field which would support and encourage the study of ecstasy, as religious founders and prophets have often had ecstatic states. But one would be wrong in this.

Notes

1. See Merriam Webster Online Dictionary, at http://www.google.com/dictionary?q=pathologize&langpair=en|en&hl=en&sa=X&ei=ps9AT-ev3KseBgAfbwMnVAg&ved=0CBIQmwMoAA.
2. Immanuel Kant, *Critique of Judgment* (New York: Cosimo Classics, 2007 (1914)), p. 87.

3. David Hume, "Of Superstition and Enthusiasm," in *Essays Moral, Political and Literary (1742–1754)*, online http://www.english.upenn.edu/~mgamer/Etexts/hume.superstition.html#anchor6.
4. Ibid.
5. Ibid.
6. Voltaire, *Philosophical Dictionary* (New York: Penguin Classics, 2004 (1764)), p. 188.
7. S. Katz, "Language, Epistemology and Mysticism," in *Mysticism and Philosophical Analysis*, ed. Steven Katz (New York: Oxford University Press, 1978), pp. 26–27.
8. Wayne Proudfoot, *Religious Experience* (Berkeley: University of California Press, 1985), p. 126.
9. Robert M. Gimello, "Mysticism in Its Contexts," in Steven Katz, *Mysticism and Religious Tradition* (Oxford University Press, 1983), p. 66.
10. Max Weber, *The Sociology of Religion*, trans. Ephraim Fischoff (Boston: Beacon Press, 1964 (1922)), p. 3.
11. Ibid., p. 158.
12. Ibid., p. 157.
13. Weber, ibid., p. 158. The Zoroastrian ceremony he describes is probably the haoma rite, whose sacred substance was often equated with the intoxicating Vedic soma plant. The original plants for both ceremonies are unknown today, thought there has been much academic guesswork about them.
14. Ibid., p. 159.
15. Ibid., p. 160.
16. Ibid., p. 180.
17. Ibid., p. 189.
18. From Max Weber, *The Protestant Ethic and the Spirit of Capitalism*, p. 53, at www.digireads.com,http://books.google.com/books?id=J2EK-m7VkeTgC&printsec=frontcover#v=onepage&q&f=false.
19. Emile Durkheim, *The Elementary Forms of the Religious Life*, p. 308.
20. Ibid., p. 327.
21. Sean McCloud, *Religious Distinctions: Class and Classification in Religious Studies and American Religion* (University of North Carolina Press). These data come from a draft of his second chapter, "The Catholic and Sectarian Depraved, the Racially Unevolved, and the Rural Degenerate: Explaining Religious Preferences in the Age of Eugenics, 1905–1934." This quote is from page 3 of the draft version.
22. Ibid., p. 4.
23. Ibid., p. 17.

24. Frederick Moran Davenport, *Primitive Traits in Religious Revivals: A Study in Mental and Social Evolution* (New York/London: The Macmillan Co., 1910), p. 27.
25. Ibid., p. 53.
26. Ibid., p. 209.
27. These included writers like Paul Popenoe and R. H. Johnson, who wrote *Applied Eugenics*; Ellsworth Huntington and Leon Whitney, authors of *The Builders of America*; and William McDougall, *Is America Safe for Democracy?* See McCloud, ibid., p. 35.
28. During the period from 1905–1934, several major books came out on this topic, especially critical of ecstatic religious revivals (as performed by "morons," inadequate," and "less-favored" people). These described a reversion to subconscious states and primitive reactions, thus moving people backward on the evolutionary ladder (evolution being a process that can move in either direction). See McCloud, ibid., p. 35.
29. Rodney Stark and William Sims Bainbridge, *The Future of Religion: Secularization, Revival and Cult Formation* (Berkeley: University of California Press, 1985), pp. 173–175.
30. E. E. Evans-Pritchard, *Theories of Primitive Religion* (London: Oxford University Press, 1965), p. 85.
31. Lucian Levy-Bruhl, *Primitive Mentality* (Boston: Beacon Press, 1966 (1923)), p. 91.
32. I. M. Lewis, *Ecstatic Religion: An Anthropological Study of Spirit Possession and Shamanism* (Harmondsworth: Penguin Books, 1975 (1971)), p. 18.
33. Ibid.
34. Ibid.
35. Lewis, ibid., p. 38. He takes his definition of ecstatic trance from the *Penguin Dictionary of Psychology.*
36. Ibid., p. 39.
37. Ibid., p. 71.
38. Ibid., p. 174.
39. Ibid.
40. Ibid., p. 175.
41. Ibid., p. 34.
42. I. M. Lewis, "Spirit Possession and Deprivation Cults," *Man* 1 (1966), p. 318.
43. William Sargant, *The Mind Possessed: A Physiology of Possession, Mysticism and Faith Healing* (New York: Penguin Books 1975 (1973)), p. 31.
44. Ibid., p. 58.
45. Ibid., p. 44.
46. Ibid., p. 45.

47. Ibid., p. 74.
48. Ibid., p. 76.
49. Ibid., p. 87.
50. Ibid., p. 86.
51. Ibid., p. 101.
52. Ibid., p. 194.
53. Richard J. Castillo, "Spirit Possession in South Asia, Dissociation or Hysteria? Part 2: Case Histories," *Culture, Medicine and Psychiatry* 18 (1994), p. 141.
54. Ibid., p. 144.
55. Castillo, ibid., p. 150. According to a 1989 study, 84% of MPD patients had at least one protector personality. We may note that the concept of multiple personality was very popular in the 1980s, with diagnosed cases reaching 50,000. After it was recategorized as dissociative disorder, its numbers declined in the late 1990s, with critical questioning about childhood sexual trauma and claims of satanic ritual abuse and the rise of malpractice suits against therapists. For a good overview of its rise and fall, see Richard J. McNally, *What is Mental Illness?* (Cambridge: Harvard University Press, 2011), pp. 139–143.
56. Castillo, ibid., p. 157.
57. This always reminds me of the Lily Tomlin line, "If you talk to God you are normal, but if God talks back to you, then you are insane…"
58. Gananath Obeyesekere, *Medusa's Hair*, p. 44.
59. Ibid., p. 91.
60. Ibid., p. 175.
61. C. Jason Throop, "On Crafting a Cultural Mind: A Comparative Assessment of Some Recent Theories of 'Internalization' in Psychological Anthropology," *Transcultural Psychiatry* 40, no. 109 (2003), p. 116.
62. Mircea Eliade is known for this affirmation of cross-cultural themes in shamanism, while Devereux emphasized specificity in different traditions. There was a debate between the two, both on whether shamanism could be considered a cross-cultural category and also whether it was considered to be a pathological or valued role in society.
63. George Devereux, "Normal and Abnormal," in *Basic Problems of Ethnopsychiatry*, trans. Basia Miller Bulati and George Devereux (Chicago: University of Chicago Press, 1980), p. 6.
64. Ibid., p. 7.
65. Ibid., p. 14.
66. Ibid., pp. 16–17.
67. Ibid., p. 25.

68. Mircea Eliade, *Shamanism: Archaic Techniques of Ecstasy* (Princeton: Princeton University Press, 1964), p. 27.
69. Grace M. Jantzen, *Power, Gender and Christian Mysticism* (Cambridge: Cambridge University Press, 1995), p. 21.
70. Ibid., p. 169.
71. Ibid., p. 222. We may note that suffering as Christ did was a major theme for both male and female mystics.
72. Ibid., p. 327.
73. Ibid., p. 331.
74. Ariel Glucklich, *Dying for Heaven: Holy Pleasure and Suicide Bombers-Why the Best Qualities of Religion are also Its Most Dangerous* (New York: HarperCollins, 2009), p. 222.
75. Ibid., p. 222.
76. Ibid., p. 224.
77. Ibid., p. 38.
78. Ibid., p. 105.
79. Amy Holly wood, *Sensible Ecstasy: Mysticism, Sexual Difference, and the Demands of History* (Chicago: University of Chicago Press, 2002), p. 280, endnote to page three.
80. Ibid., p. 8.
81. Ibid., p. 19.
82. Ibid., p. 20.
83. Ibid., p. 240.
84. Ibid., p. 239.
85. Ibid., p. 256.
86. Rudolph M. Bell, *Holy Anorexia* (Chicago: University of Chicago Press, 1985), p. 16.
87. Ibid., p. 13.
88. Ibid., p. 10.
89. Ibid., p. 21.
90. Ibid., p. 56.
91. Ibid., p. 113.
92. This controversy was strong enough to cause Kripal to leave the study of Indian religion entirely and move into a different field of study.
93. Jeffrey J. Kripal, *Kali's Child: The Mystical and the Erotic in the Life and Teachings of Ramakrishna* (Chicago and London: University of Chicago Press, 1995), p. 2.
94. Ibid., p. 121.
95. Ibid., p. 292. A major problem in the book is that Kripal took Bengali idioms and jokes literally: When Ramakrishna made a joke about dysentery, Kripal interpreted this literally, saying that he is referring the religious ecstasies as intestinal problems.

96. Ibid., p. 303.
97. Ibid., p. 299. Or as he phrases it that something more than Vedantic instruction was going on in that hut.
98. Ibid., p. 320.
99. Ibid., p. 32.
100. Ibid., p. 327. However, we should note that Kripal was not trained as a psychoanalyst.
101. Sigmund Freud, *The Future of an Illusion*, trans. W. D. Robson-Scott and ed. James Strachey (Garden City: Anchor Books, Doubleday and Co., Inc. 1964 (1927)), p. 3.
102. Sigmund Freud, *Civilization and Its Discontents*, standard ed. 21, p. 64, cited in William B. Parsons, *The Enigma of the Oceanic Feeling* (New York: Oxford University Press, 1999), p. 35.
103. Parsons, ibid., p. 97.
104. Ibid., p. 98.
105. Ibid., p. 108.
106. The drives toward life and death are discussed in Sigmund Freud's *Beyond the Pleasure Principle.*
107. J. Moussaieff Masson, *The Oceanic Feeling: The Origins of Religious Sentiment in India* (Dordrecht: D. Reidel Publishing Company, 1980), p. 4.
108. Ibid., p. 60.
109. Ibid., p. 70.
110. Ibid., p. 75.
111. M. Moussaieff Masson, "The Psychology of the Ascetic," *Journal of Asian Studies* XXXV, no. 4 (August 1976), p. 613.
112. Ibid., p. 622.
113. Ibid., p. 623.
114. Bertram Lewin, *The Psychoanalysis of Elation* (New York: W.W. Norton and Co., Inc. 1950), p. 14.
115. Ibid., p. 15.
116. Ibid., p. 60.
117. Ibid., p. 129.
118. Ibid., p. 70.
119. Ibid., p. 137.
120. Ibid.
121. Ibid., p. 138.
122. Ibid., p. 143.
123. Ibid., p. 149.
124. Ibid., p. 152.
125. Ibid., p. 153.

126. Ibid., p. 156.
127. Ibid., p. 157.
128. Ibid., p. 173.
129. Russell Shorto, *Saints and Madmen: Psychiatry Opens Its Doors to Religion* (New York: Henry Holt and Company, 1999), p. 13.
130. Cited in Shorto, ibid., p. 66.
131. D. Raybeck, J. Shoobe, and J. Grauberger, "Women, Stress and Participation in Possession Cults: A Reexamination of the Calcium Deficiency Hypothesis," *Medical Anthropology Quarterly* 3, no. 2 (1989), pp. 139–161.
132. Alice Kehoe and Dody H. Giletti, "Women's Preponderance in Possession Cults: The Calcium-Deficiency Hypothesis Extended," *American Anthropologist* 83, no. 3 (1981), p. 550.
133. Julian Jaynes, *The Origin of Consciousness in the Breakdown of the Bi-Cameral Mind* (Mariner Books, 2000).
134. John Horgan, *Rational Mysticism: Dispatches from the Border Between Science and Spirituality* (Boston: Houghton Mifflin Company, 1983), p. 91.
135. Ibid., p. 91.
136. Ibid., p. 93.
137. Ibid., p. 95.
138. Shorto, ibid., p. 186.
139. Personal communication, Charleston, 2005.
140. Ethan Watters, *Crazy Like Us: The Globalization of the American Psyche* (New York: Free Press, 2010), p. 2.
141. Ibid., p. 7.
142. Ibid., p. 33.
143. Ibid., p. 109.
144. Ibid., p. 107.
145. Ibid., p. 135.
146. Ibid., p. 137.
147. Richard J. McNally, *What Is Mental Illness?* (Cambridge, MA: Harvard University Press, 2011), 26.
148. Ibid., p. 1.
149. Ibid., p. 39.
150. See https://www.huffingtonpost.com/2013/05/17/dsm-5-unveiled-changes-disorders-_n_3290212.html. However, there is some debate over disorders vs behavioral problems and behaviors.
151. For a discussion of these issues, see my book *The Madness of the Saints: Ecstatic Religion in Bengal* (Chicago: University of Chicago Press, 1989).

152. Watters, ibid., p. 156.
153. This list comes from Derek Summerfield, cited in Watters, p. 65.
154. More recent interesting pathologies include micromastia (the disease of women having breasts that are too small for the cultural ideal) and female sexual dysfunction (when women want sex less often than they should want it).

CHAPTER 4

Attacks on Ecstasy: Theology—We Don't Want It Either

The academic approach to religion tends to shunt the study of ecstasy into the theological camp. A major insult, from the academic perspective, is to say that any writer appears to view ecstatic states with some sympathy is a theologian, or possibly worse, a crypto-theologian.[1] However, all three of the great Western religions have had reasons for rejecting ecstatic states, especially in their mainstream forms. They are often no more sympathetic toward religious ecstasy than the critical theorists on the other side of the aisle, accepting a narrow range of experiences within their own systems of belief and practice, and rejecting ecstasies which occur outside of their own systems. We shall examine some of the approaches to ecstasy in Western Theology here.

Judaism

In Judaism, the major source for the study of ecstatic states would be the Biblical prophets, with a secondary focus on Kabbalistic writers. The Biblical prophets claimed to have direct contact with God, through ecstatic visions and auditions, and acted as messengers or intermediaries. Prophets were formal representatives, and their messages conformed to the teachings of the Torah. The *navi* or *nabi* could speak God's word and act as the mouth of God. Prophecy frequently caused fainting, temporary insanity, involuntary muscular spasms, and seizures; some prophets were capable of receiving prophecies in their sleep.

J. McDaniel, *Lost Ecstasy*, Interdisciplinary Approaches to the Study of Mysticism, https://doi.org/10.1007/978-3-319-92771-8_4

The Talmud counts fifty-five "historical" prophets whose prophecies were recorded in the Hebrew Bible because they contained a message relevant to all generations. These included forty-eight male prophets and seven female ones.[2] Prophetesses included Sarah (wife of Abraham), Miriam (sister of Moses), Deborah (the only woman among the "Judges"), Chana, Avigayil, Chuldah, and Esther. There were fifteen prophets whose words were recorded in individual books that bear their names: Isaiah, Jeremiah, Ezekiel, and twelve lesser ones including Amos, Hosea, Nahum, and others. The other forty are mentioned in various places throughout Bible (an example would be Nathan, in the Books of Samuel). There were also unrecorded prophetic experiences mentioned.

The era of prophecy officially came to an end about 2300 years ago. The last generation of prophets were those who began to prophecy before the First Temple was destroyed in 423 BCE, though a number of that generation survived the 70-year Babylonian exile and lived to see the building of the Second Temple. Most famously, Ezekiel prophesied in Babylon, and three prophets, Haggai, Zachariah, and Malachi, were members of the "Great Assembly" that led the people in the first years of the return from Babylon. Mordechai and Esther were also members of the long-lived generation that mourned the destruction of the First Temple and witnessed the building of the second. With the demise of that generation, "prophecy departed from Israel."

As the Babylonian Talmud's commentary phrases it, "when the former prophets died, the Urim and Tummim were cancelled."[3] These former prophets, including Haggai and Zechariah, were considered to be still alive at the fall of the Temple. There would be no further holy writings beyond the Hebrew canon at the time, in which it would be alleged, "God spoke to me, saying…"[4] The Urim and Tummim (or Thummim) were part of the breastplate worn by the high priest of the Jerusalem Temple. The terms refer to lights and perfections, thus symbolically revelation and truth. The high priest of the Temple had been able to communicate with God, and receive divine guidance, but his power was taken away.

There would be no more direct communication with God after the fall of the Temple, though indirect communication was allowed, the "echo" of relationship by law, and by the correct interpretation of scripture. As Jacob Neusner states, "the study of Torah for sages has now replaced prophecy."[5] The ending of prophetic inspiration brought the rise of rabbinic Judaism as the mainstream form of Jewish tradition.

The study of prophecy has become a historical study—there are no legitimate Jewish prophets in the modern world. Biblical study includes both accepted prophets and also "false prophets." There have always been doubts about prophetic inspiration. There are two types of "false prophet" described in the Hebrew Bible: the one who claims to be a prophet in the name of a false god (thus idolatry), and the one who claims to be a prophet in the name of the God of Israel, but declares that any word or commandment which God has said no longer applies, or makes false statements in the name of God. As traditional Judaism believes that God's word is true eternally, one who claims to speak in God's name but diverges in any way from what God himself has said, cannot be inspired by divine authority. As Deuteronomy 13:1 states, "Be careful to observe only that which I enjoin upon you; neither add to it nor take away from it." A true prophet must be able to focus all of his or her faculties on God. Such experiences were transcendent, lifting the individual out of the ordinary human frame of reference by feelings, visions, and other phenomena and are often referred to by analysts as ecstatic. As one commentator remarked: ecstasy is the focusing or concentration of the faculties upon God.[6] He speaks for the correct god and agrees with everything that he has commanded. The words of a legitimate prophet always come true.[7]

Some sources suggest that the loss of prophecy was punishment for sin. Over two hundred years before Malachi, the prophet Amos predicted the cessation of prophecy: "A time is coming—declares my Lord God—when I will send a famine upon the land: not a hunger for bread or a thirst for water, but for hearing the words of the Lord. Men shall wander from sea to sea and from north to east to seek the word of the Lord, but they shall not find it" (Amos 8:11–12). A Midrashic commentary states that many Jews failed to return to Israel after Cyrus gave them permission and therefore prophecy ceased.[8]

Another theory about the ending of prophecy was based on the destruction of the Temple. Ezekiel Chapters 8–10 describe a vision in which God shows the prophet the idolatry in Jerusalem, and God's presence abandons the Temple and goes into exile. Thus, divine anger over idolatry contributed to the disappearance of prophecy. Another theory bases the ending of prophecy on the lessening of idolatry. This idea meshes with a Talmudic statement that at the beginning of the Second Temple period, the temptation for idolatry ceased being the force it had been during the First Temple period. Rabbi Yehudah HeHasid argued that once the urge for idolatry vanished there no longer existed the need for prophecy to counterbalance magic.[9] With less desire for idolatry, prophetic revelation would have too much power if left unchecked. To preserve free will, prophecy had to cease as well.[10]

We later see a return to mystical and ecstatic states with Hasidism and the various understandings of Kabbalah. But in mainstream forms of Judaism, the focus is rabbinic and emphasizes the law. Individual ecstatic states have been discouraged in favor of textual interpretation since the time of the prophets, for Judaism's response to the ending of prophetic inspiration has been the elevation of law and tradition. We see the rise of ecstatic and mystical Judaism in the medieval period with the Zohar, with Abulafia's "prophetic kabbalah" and the heavenly journeys and miracles of Hasidic Judaism. However, these have not gained wide acceptance and remain minority traditions. The arguments of rabbinic Judaism against Kabbalism have variously accused it of dualism, paganism, lack of mention in the Talmud, anachronism, and violation of the commandments—they have remained in force since the time of the Gaon of Vilna's attacks on the Hasidim.

Today's dominant understanding of ecstatic experience in Judaism follows the approach of J. B. Soloveichik, a major thinker in Orthodox Judaism. He emphasized the sober, painstaking, unemotional approach found in halakhic Judaism. The greatest good and highest privilege is obeying God's will as it is revealed in Jewish law, while religious ecstasy is viewed with suspicion, as unnecessary and unreliable. The only reliable way to understand God is through the law, not through ecstasy. The prophets are gone, and ecstasy is not to be trusted.

Islam

We may note that in mainstream forms of Islam, the ending of prophecy is even clearer. The Qur'an mentions twenty-five prophets by name, though there are Hadith or traditions which give a higher number (124,000, according to Ibn Hanbal). But it is unambiguous that

Muhammad is the last of the prophets, the seal of the prophets, who came for all of humanity, as stated in the Qur'an (16.36). To deny this *sura* or stanza, and claim to be a prophet after Muhammad, is to commit heresy (thus, the problems seen in Islam's relation to the Bahais and the Ahmadiyyahs, among other groups, who claim later prophets). Ecstatic states are too close to prophetic ones, largely forbidden, though overlooked in some countries which allow the practice of mystical Sufi traditions.

Today, the legality of Sufism shifts, according to the strictness of the country's government and powers of its religious factions. In Sufism, Mohammad is not only a historical prophet, but also a role model. His ascent to heaven, the *miraj*, is a form of ecstatic journey that later followers can perform as well. Allah is not merely a distant judge and war commander; he is someone to love passionately. The schools of Sufism emphasize love and union in various ways, but all see closeness to Allah as the goal of Islam.

Most mainstream Muslims regard Sufism as non-Islamic, a kind of heresy, based on two grounds. First, they regard the Sufi desire to let go of ego and experience Allah as All to be a form of *shirk*—the sin making oneself like God. Second, Sufis honor great Sufi saints, venerating their grave sites and going there on pilgrimages. Mainstream Islam, again, regards this as *shirk*, worshipping human beings instead of God, and graves and tombs have been destroyed in some Islamic countries. There is also the concern that Sufism is never found in the Qur'an, in the prophetic traditions, the early years of Islam or in the recorded wordings of the Companions. It is thus interpreted as an innovation forged to misrepresent Islam and twist its teachings to serve personal whims and desires.

Sufis have been especially objectionable to members of the modern Islamic State group. They call them *mushrikin*, an Arabic word meaning polytheists. There have been recent attacks on Sufis in several countries, including desecration of the shrine of the Sufi philosopher Lal Shahbaz Qalandar. Sufis had come bearing offerings of rose petals and had prayed at the tomb of the saint, and hard-liners saw their faith as an affront to Islam, which holds that there is a single, indivisible God. Since at least 2016, Islamic State militants have targeted Sufis, and systematically razed the tombs of Sufi saints and dynamited their shrines. Recently, the Islamic State began carrying out mass executions of Sufi worshipers during prayer.[11] Their publications make clear that it considered Sufism to be one of the main "diseases" of Islam, and there were areas where it planned to "eradicate" Sufi beliefs.

Sufism has been and continues to be forbidden in many Muslim countries, for its emphasis on religious ecstasy makes it suspect, heterodox rather than orthodox. But the major Muslim response to the ending of prophetic inspiration has also been textual, to develop a legal system which can interpret the Qur'an and Sunnah traditions in the modern world.

Christian Theology

Ecstatic states have been under suspicion in Christianity since its early formation as an institutional religion. In the fourth century, when the Bible was canonized and church dogmatic Theology was being organized, St. Augustine wrote that the miracles and healings described in the New Testament were not allowed to last into modern times and were only needed to authenticate Jesus' teachings and convince the multitudes. The time of signs and wonders was over.[12] Though later in his life, Augustine came to reject the ending of miraculous events and placed a greater emphasis on religious experience, his early writings were more influential for the development of church teachings.

This idea of the "closing of the canon" and the ending of the sorts of experiences described in it came to be called Cessationism, the ending of supernatural events after Biblical times. The idea became popular among many early Christian writers. Catholic doctrine rejected complete Cessationism—it just emphasized the rarity of supernatural events and the need for evidence of their validity. This became important because of the medieval emphasis on the miraculous lives of saints, martyrs, and heroes. These exemplary figures were associated with stigmata, levitation, miraculous healings, powerful relics, ecstatic visions of heaven—and sometimes claims of new doctrines as legitimate. The Catholic Church did not reject all apparently supernatural events, but said that many were the result of imagination, error, deceit, and illusions created by Satan. One needed both historical verification and comparative Theology to discern the accuracy of such claims.

Ecstatic and supernatural states were also highly suspect for the theologians of the Protestant Reformation. Such states came to be associated with the myth and superstition that they believed to be a part of Catholicism. Both Martin Luther and John Calvin argued in favor of

Cessationism. Calvin used it to discredit the Catholic mystical tradition, which he felt had too much emphasis on unvalidated experience. He denied the continued existence of the "spiritual gifts" described by Paul and accepted the five senses and reason as the only reliable way to know truth (which was to understand scripture and learn God's will through it). The only legitimate religious experience was conversion (revelatory dreams and visions were only for people in the Bible), and all other religious experience came from the Devil.[13]

Thus, the total focus of Calvinist (and later many other forms of Protestant) experience came to be conversion, and witnessing to others about it. Otherwise, any religious experience was tainted by natural depravity, for in Calvin's view all human beings were born depraved and affected by original sin. Even conversion experience must fit into narrow doctrinal and ritual confines, the model of the proper conviction-salvation experience. Otherwise, the person went beyond the Biblical evidence (and acceptable interpretations of the Bible). There are no analogous spiritual events in the modern world to those in the Bible, and faith must overcome depravity without modern evidence or ecstatic experience. After Cessationism and the Protestant rejection of the supernatural, the only possible interpretations of ecstatic religious experience beyond specific doctrinal limits were as events of demonic origin, or as errors.

Catholic and Protestant theologians have had different styles of attack on false ecstasy. Medieval Catholic Theology accepted the idea of demonic possession, and there was great interest in demonic simulations and counterfeits, the Devil's illusions and lies. The person could not trust his or her own senses, as the flesh is the realm of the Devil, and only intellectual visions were legitimate, as they had no sensory content. In Protestant Theology, there was a drama of sin and salvation, and human sinners were not considered worthy of seeing God and would not do so until the Last Days. The emphasis was on sin rather than demons. This has changed with the rise of the Charismatic movement, in which ecstatic states and demons have returned to the modern world.

An extended overview of understandings of ecstasy in Catholic Theology may be found in Albert Farges' book *Mystical Phenomena Compared with their Human and Diabolical Counterfeits*.[14] He contrasts legitimate and illegitimate forms and interpretations of ecstatic and mystical experiences, primarily following the ideas of St Teresa of Avila and St. Thomas Aquinas. He calls mystical Theology a "true science," a collection of facts and theories based on experience.[15] It is also "queen of the sciences" as a form of Theology, for it allows mankind to communicate with and gain knowledge of God, while maintaining reason.

While striving toward perfection, the soul may experience such gratuitous gifts of God as ecstasy, rapture, visions, and prophecy. These are infused by God into the soul by grace and may be understood as part of an active or passive spiritual life. However, the deepest states are primarily passive, with God bestowing supernatural phenomena which people cannot attain on their own.[16] He contrasts the inferior or ordinary degrees of sanctity with superior ones (such as ecstasy and stigmata) and notes that it is important for the church to be able to distinguish between types of mystical phenomena. This is because of the "dangerous wave of false mysticism of the occult and spiritist sciences," and the similarity between the claims of mystics and those of hysterics in mental hospitals.[17] The supernatural must not be confused with the marvelous, whether through ignorance or malice. Otherwise, indiscreet and intemperate people will desire such experiences, without realizing that they are unworthy, and naïve and imaginative people might imagine that they already have had such experiences, as some people who read medical books imagine that they have morbid symptoms.[18]

"True" ecstatic and mystical states were found in people like Jesus, St. Paul, and the early martyrs. Such states were shown by accurate prophecies in visions, and indifference to pain and death. Early hermits and anchorites would go into extended ecstatic states which could last all day or all night, and included symbolic and prophetic visions, gifts of healing and power over nature. Medieval ecstatic saints had visions of light and left symbolic dialogues and treatises. Such states involve an infusion of sanctifying grace from God, a direct intuition or "sensation of the divine."[19] God's presence evokes love, peace, humility, heroic patience, and a desire for justice, and the ordinary powers of the soul are suspended, while the spiritual senses (such as supernatural sight and hearing) are evoked. The person is transformed by holiness.

However, as Farges points out, there are other simulations of these states, which come from illusion, delirium, hypnosis, madness, physiological problems, and abnormal excitement of the imagination, as well as exploitation by charlatans. "True" experiences include mystical jubilation and spiritual inebriation, tremors, transports, exaltation, spiritual sleep, estrangement of the ordinary senses and working of the spiritual senses.[20] Ecstasy is "the flowing of the mind and soul into God, as the ocean of divine life," while rapture is sudden and impetuous, the flight of the spirit which separates the soul from the body.[21] On the other hand, illegitimate or false or counterfeit ecstasy would include pride, desire, restlessness, lethargy, and the desire to be observed. Trying to bring about ecstatic states through one's own efforts is a sign of a false state, as is living in a state of sin, incoherent words and scandalous actions, and having no memory of the ecstatic state.

Farges notes that these false states may come about through the Devil's actions, or simply through human sin and illness. The Devil cannot actually possess a good person (God will not let people be possessed by the Devil whom he wants to possess himself), but the Devil is allowed to tempt people and cause obsession.[22] Satan can give them fascination with repulsive forms of gluttony, blasphemy, and sexuality, making them attracted to objects of greatest horror, and create vivid visions of sin. His presence may also be found in strange diseases which appear and disappear, feelings of pain and ugliness, and the hatred of God.[23]

The Devil can only create certain types of experiences. The intellectual vision, the sense of divine presence which is beyond sensation and imagery, is beyond the Devil's powers to simulate. This involves the deepest level of the person, the apex or highest sphere of the spirit, where only God can communicate ideas. However, the lower areas of the person, the spirit (the area of reason, will, and love, the realm of the imaginative vision), and the soul (the area of sensation, appetite, and passion, where visions may appear as external) can be reached by the Devil's ability to deceive. Hallucinations can be caused by the Devil, and they may bring coarse and ridiculous imagery. But the Devil cannot touch the topics of legitimate ecstatic vision: God, Jesus Christ, Virgin Mary, the saints and angels, and the souls in Purgatory.[24] The imagery that the Devil causes or motivates is incomplete or imperfect; diabolical intervention has limitations:

1. The Devil cannot create new images; he can only recycle old ones. He cannot make a person blind from birth see color.
2. His imagery involves subtle malice, is contrary to Catholic faith and morals, and is exaggerated, deformed, and corrupt.
3. Diabolical intervention causes emotional effects, and the person ends up feeling anxious, fearful, sad, disgusted, and proud.
4. The Devil cannot corrupt the virtuous and faithful, but his influence can be seen in the sinful, vain, and credulous.[25]

The hallucinatory counterfeit created by the Devil is a "false outward reflection of an inward image," projected into the outer world, and thought to be sensory.[26] There are tests which can determine whether or not an image is of diabolical origin. There is real presence of an object if the eyes are closed or the back is turned, and the vision ceases. However, a subjective image is independent of external events and can be seen with the eyes shut, in every direction, and one cannot turn one's back on it. True vision does not move with the eyes and appears to be stationary, while hallucination changes and can be superimposed on other images. Such hallucinatory images may come from the Devil, or from emotional states, poisonous agents, and natural pathological causes.[27] There are also perceptual illusions (false interpretations of real data), waking dreams, internal fictions, and temporary aberrations of mind, especially in the areas of imagination and reason.

The ideal situation for evaluating an ecstatic state is to catch a mystic during the vision, for then the vision itself can be tested. The observer may place a screen between the ecstatic and the physical locale of the vision, and if the vision is cut off by the screen, then it is a true vision. The image can be checked for transparency, for the source of illumination, for distortions of space, weight, and diffusion of sound and light. True supernatural visions are marked by their respect for the laws of normal sensory reality and show harmony and order.[28] There are also such indirect proofs as miracles and levitation. One may also observe the life of the ecstatic: if there is humility and virtuous action, then the event is more likely to be true, while if the person desires to be set apart from others or indulges in vain display, the vision is more likely to be false. Diabolical counterfeits of ecstatic visions are described as clumsy, with grotesque, indecent, hideous, or malicious images.

For Farges, true vision can only occur within the Catholic context. He does mention natural visions, which are illusory but not demonic, but all visions which claim to be supernatural and do not follow Catholic teachings are false. He thus negates the legitimacy of most of the religious and ecstatic visions in the world. However, he is quite indignant that philosophers and psychologists have denied the truth of Catholic ecstatic states. As he notes:

> During the whole period of its history until the present time, the Catholic Church has constantly regarded ecstasy- we speak of true ecstasy, the ecstasy of the saints- as something supernatural, sacred, and worthy of all

> respect from us… now this marvelous phenomenon of divine ecstasy… has been denied, travestied, and odiously scoffed at by a crowd of contemporary rationalists, scholars and philosophers.[29]

These materialistic scholars, physiologists, and physicians "see in ecstasy nothing but a pathological case of neurosis bordering on madness."[30] It is called a degeneration of normal faculties, mental aberration, and a disorder of the nervous system. Such critics think that ecstasy is brought about by too much contemplation of religious ideas, or in the bubblings of the subconscious. Some even equate it with immorality, as one writer phrased it, "love of God is a purely sexual emotion."[31] They claim that "ecstasy suppresses reason and liberty in man" leaving only sentiment, thus "narrowing the field of consciousness."[32] Such a view suggests that the interior life of mankind is empty and barren, and that without ordinary sensory ideas there can be nothing of value there. He also objects to loose usage of the term "ecstasy," as in the "capitalist who calculates some operation which will raise him to the highest summit of fortune." This is not really ecstasy, for objects cannot generate that state. "No object in this world known to human reason… is capable of producing alienation of the senses or bringing about ecstasy."[33] Only mystical contemplation of God can bring about true ecstatic states.

Farges follows St. Thomas Aquinas, who divides ecstasy into natural, diabolical, and divine types. Natural or non-Christian mystics may have imitations of true, theistic mysticism, due to human efforts (with formulas, recipes, dancing, or chanting) but these are not true ecstasy. His tolerance for non-Christian religious states is limited: "How have serious minds been able to imagine that it would be possible to attain to God by such purely mechanical and ridiculous means? They only lead to nothingness, or the Buddhist nirvana."[34] These, we can see, are not desirable states. Only the proper form of ecstasy gives the true vision of God. Catholic Theology accepts a very limited version of ecstasy, which must conform to both mystical and ethical theological norms. It thus rejects an enormous amount of experience—and began the movement of Scholastic pathologizing of ecstatic states in the West.

A critique of ecstatic states from the Protestant perspective is found in R.A. Knox's well-known *Enthusiasm: A Chapter in the History of Religion*. While the modern definition of enthusiasm emphasizes appreciation and enjoyment, older definitions are more specifically religious. Enthusiasm is "*en-theos*," god within, an "eccentric spirituality" in which "an excess of charity threatens unity." As Knox notes, it should really be called "ultrasupernaturalism" for the person who "expects more evident results from the grace of God than we others."[35] The enthusiast seeks angelic purity, alternating between rigorism and antinomianism. "The emphasis lies on a direct personal access to the Author of our salvation," and for Knox, this is anthropocentric, more concerned with man than with God.[36] Though Knox states that he does not wish to criticize enthusiasm, he spends several hundred pages doing so.

Knox argues that enthusiasts are elitist, seeking theocracies in which they are leaders, thus motivating schisms, heresies, and rebellions. He notes a symptom of enthusiasm is ecstasy, "a mass of abnormal phenomena," which includes "unintelligible utterance," trance, convulsive movements, and prophecy. It is difficult for the observer to tell if it is hysteria or diabolism, "but certainly it is puzzling."[37] Such "freaks of religious history" took shape after Luther, with Quakers, Pietists, Camisards, Shakers, and Methodists. While he finds most enthusiasm among Reformation Protestants, it could also be found in Catholic Quietism and Jansenism. He notes the origin of the word's popularity in Bishop Hickes' sermon on "The Spirit of Enthusiasm Exorcised."[38]

Knox is also concerned that "spiritual manifestations" are forms of idolatry that bring the sacraments into disrepute.[39] He wonders if it is the air or the soil that induces such states as "shriekings, roarings, groanings, gnashings, yellings, cursings, blasphemies and despairings." In the Methodist case, he criticizes Wesley as credulous and writes of revival participants that "Anybody but Wesley was content to write them down as lunatics."[40]

After a good deal of condemnation, Knox does create a model, contrasting the mystical and evangelical strains of enthusiasm. The mystical form is quiet and contemplative, while the evangelical type is focused on sin, redemption, and religious self-expression.[41] Either tendency can be "a signpost to the morass." While the mystic divorces soul and body, and is "tone-deaf to theology," the evangelical "children of predestination" consider themselves to be emancipated from the bondage of law, beyond concern for sin because they are already saved, inherently different from the ungodly, who have little worth. Both forms of ultrasupernaturalism are aberrations, distrusting human reason in favor of inner conviction.[42]

Knox relates ecstasy to hysteria and diabolism, and quotes St. Paul on the dangers of unregulated spiritual gifts. He finds ecstasy to be of greatest danger to women: "From the Montanist movement onwards, the history of enthusiasm is largely a history of female emancipation, and it is not a reassuring one."[43] The author felt that the ecstatic states of female enthusiasts threatened society, through female teachers and "convulsionary priestesses," whose "religious disorders" showed "an unfettered exercise of the prophetic ministry," which "can threaten the ordinary decencies of ecclesiastical order."[44] From the Montanists, who began in Phrygia (the "land of heathen orgies") came an "atmosphere of ecstaticism" which has run throughout the history of Christianity. These have been dangerous to the Christian tradition, which he felt to be a tradition of rationality and order. While Knox mentions several early debates about the role of ecstatic states in Christianity, he determines that the only good ecstatic is a silent one. He admits that early Christian prophets may have gone into ecstatic states, as St. Teresa did, but at least they did not speak while in the state of ecstasy—they described their experiences afterward. This differentiates the acceptable early Christian groups from the charismatic ones, which bring disorder to the church.[45] For Knox, ecstatic states show vanity, eccentricity, and abnormality, thus he speaks of them as religious "freaks and aberrations."

There are more modern sources for both Catholic and Protestant critiques of ecstatic states. Perhaps the most modern and popular Catholic theological book on the analysis of ecstatic visions is Fr. Benedict J. Groeschel's 1993 book, *A Still, Small Voice: A Practical Guide on Reported Revelations.* Its focus is not ecstatic pathology, but rather ecstatic unimportance. Ecstasy is something to be avoided, not because it is a disease or a symptom of insanity, but rather because it is irrelevant. Its motto on the title page, and mentioned through the book, is St. Therese of Lisieux's statement, "To ecstasy, I prefer the monotony of sacrifice." Groeschel has a dry wit in his description of the attitudes of his fellow theologians toward ecstatic events:

> Reports of weeping madonnas, miraculous physical cures, prophecies and heavenly warnings generally leave members of the clergy annoyed, perplexed and skeptical... Because these events rarely happen to the clergy, who are terrified of being seen as superstitious or credulous, the average priest or minister will plead ignorance of such things, while rabbis will be grateful that these events are almost unknown in the Jewish world outside of Hasidism. Many of the clergy hoped and thought that such religious phenomena belonged to another time and were glad to leave them there.[46]

He does not speak of their potential Satanic evil, but rather he emphasizes the trivial nature of ecstatic experiences, as they are private revelations which do not have significance for the church. He prefers for people to focus on the non-visionary saints, known for their morality and service to the church: "It is important to keep clearly in mind that private revelations have no significance apart from the public relation of Sacred Scripture, interpreted by the traditional teaching of the Church."[47] Any new revelations must be only for the good of the people involved, as Pope John Paul II noted in 1983 that "no new public revelation is to be expected" before the return of Christ. On some occasions, church evaluations can be wrong, as they were with St. Joan of Arc, who was condemned as a witch by a church court, and later exonerated and canonized. However, they are more trustworthy than private revelations, whose content is often esoteric or trite, showing desire for attention and power, on the part of a histrionic personality or paranoid schizophrenic. There are also religious charlatans, who can be a spiritual menace.[48]

One major problem with ecstatics is that they take symbolic visions literally, and these range from angry revelations to saccharine ones. The angry revelations, which are of course false ones, condemn the pope and the church, with bishops and the Catholic hierarchy on their way to hell. As Groeschel phrases it, "the Lord is coming, and he is coming with an axe."[49] There are dire warnings and false prophecies. The more saccharine types also have an antiauthoritarian bias, especially those by the "mawkish disciples of Mother Earth." They are followed by people addicted to revelations, often fearful or seeking after magical power.

When Groeschel examines ecstatic visions, he has several questions. Is the revelation contrary to Catholic faith? Is it absurd or confused? Is it unoriginal or derivative, a rehash of existing doctrine? Is it inconsistent? Does the ecstatic appear defiant, proud, judgmental, mentally ill, involved with magic or the occult? Thus, the ecstatic vision must tread a fine line, fitting into a narrow range of acceptability. If it is too close to Biblical teachings, it is disqualified by being derivative. If it is too far from them, it is disqualified by being outside of church teachings. In either case, it is unimportant for the church as a whole, which should properly depend on scripture, tradition, and reason.

One major distinction between Catholic and Protestant approaches to ecstasy is based on categories of knowledge. The modern Protestant understanding of ecstasy collapses the traditional tripartite division of ecstasy (diabolical, natural, and supernatural) into a dualism—God or truth versus Satan or falsehood. The old category of the natural ecstasy is ignored and generally collapsed into the diabolical: Thus, Eastern and New Age religions become equated with Satanism. Even deep breathing meditation for the sake of relieving stress becomes dangerous—for if God is not actively in the mind, Satan will be there. Because there is no category of natural ecstasy, Satan gains much greater ground. There are modern Protestant movements which have developed specifically to counter such evil ecstatic states. One of these emphasizes deliverance from evil states and beings.

The Deliverance Movement is an approach to problematic intense and ecstatic states found in popular charismatic and non-denominational Christianity. It is based on the idea that the Devil and his demons are still active in the world, and that human beings must fight with them. This belief follows a literal reading of the New Testament, in which we not only have gifts of the Holy Spirit and miracles, but there are also descriptions of Jesus Christ casting demons out of people and apostles casting them out in Jesus' name. Ecstatic religious experience, in this context, can be interpreted in only two ways—it is from Jesus or from Satan. There are no further interpretations possible. Thus, the origin of the phenomenon must be determined, and if it is unacceptable, a remedy must be found. While the concept of spiritual warfare was part of the early church, the Deliverance Movement developed in the 1970s, with televangelists like Benny Hinn, the rise of dispensational Christianity, and the "third wave of the Holy Spirit" with signs and wonders.

Deliverance Theology includes a Protestant version of the Catholic practice of spiritual discernment, which is necessary for spiritual direction of believers and also for exorcism of demons. The major modern analyses of ecstatic states, in both mainstream Catholicism and Protestantism, are largely psychological. Ecstatics in these traditions might have problems in interpreting what is happening to them, or have childhood traumas, or medical problems, but they rarely have demons. However, Deliverance ministers emphasize the role of demons, and if so, which ones.

In Deliverance thought, any intense passions toward sexuality and violence (as well as other actions considered to be sinful) are understood to result from demonic interference. Demons are loosely equated with evil spirits, ghosts, ancestors, and foreign gods. They are variously called unclean, distressing, perverse, and of ill will, and they draw people toward destructive and transgressive passions. They are removed by exorcisms which cast out Satan and his minions, often at Christian revivals and crusades. People fall writhing to the ground, kicking and screaming, as the demons manifest. Many revivals have "stretcher-bearers" who take such people to a special Deliverance tent for prayer, exorcism, and "power encounters."

Exorcism is democratized, no longer confined to a priestly hierarchy. In a book popular in many charismatic circles, *How to Cast Out Demons: A Beginner's Guide*, Doris. M. Wagner describes ways to discern and respond to demonic possession.[50] The victimized person is "afflicted," "in bondage," "tormented," and "oppressed." He or she undergoes a transformation, has an urge or hears a voice saying to do something sinful, and is compelled to act. The person is unable or unwilling to stop, and immune to prayer and confessions. This may be aggravated by past involvement with other religions, especially Freemasonry.[51] Demons are revealed in emotional problems, especially those of violence and anger, but also man-hating, frigidity, and attraction to pornography.[52] Demons are associated with a whole range of problems. There are demons of skepticism, criticism, and unbelief (especially found in academics) and demons of apathy, allergies, financial problems, and illnesses. She gives a list of demonic fears, which includes fear of failure, the dark, violence, the future, heights, water, spiders, snakes, men, women, open spaces, public speaking, flying in airplanes, grocery stores, and loud noises.[53] There are even evil spirits of yoga.

Demons may be carried down in families, or they may attack by contagion. There are inherited spirits, called familial or generational spirits, which create problems through the family lines. We find spirits of lust (from ancestors who had affairs), of addiction and depression and rejection, and spirits of the New Age (due to ancestors who practiced witchcraft).[54] Demons which attack by contagion are rather like germs—one can catch them from infected people, or from infected objects. People also catch demonic contagion from other people during intense, traumatic experiences, and create a "soul tie" or link between people down which demons may travel. The demon is thus transferred from one victim to another by this means. The link must be severed via exorcism, or it can become a channel for multiple demons to enter.[55]

Spirits may dwell in objects (especially statues and totem poles), also paintings, and photographs.[56] They may also spend time waiting to pounce in demonic video games and music. Indeed, even a peaceful state of mind is dangerous, for in meditative or inactive mental states demons may also come to possess people. The healer must identify the demonic entry points, discern the evil spirits, and get rid of them. Demons react intensely when being extracted from their victims, so such victims must be "faith-shielded," "blood-protected," and "blood-cleansed," and must wear the armor of spiritual warfare.[57]

The belief in spirits living in objects shows a sort of Christian animism, in which spirits may inhabit the material world, with varying degrees of power and influence. Deliverance Theology is a modern throwback to earlier forms of belief, and the majority of ecstatic states it describes are considered to be abnormal and evil. As in older Catholic and Protestant theologies, ecstasy is dangerous, and likely to have evil origins.

While there are no standard textbooks in Protestant thought on distinguishing types of ecstasy, there have been theologians interested in the topic, and some have been sympathetic to the loss of ecstatic states. The Presbyterian minister and psychologist James Loder states that theologians have problems with religious ecstasy,

> Yet often theology, which should be the language of conviction, has also had trouble with these experiences. They are too subjective for those preoccupied with rigorous demands for theological thinking, and more unique and particularized than can be coped with by highly generalized theological systems. It is this tendency in much of modern theology to discredit convictional experiences that is partly responsible for their repression among both clergy and laity.[58]

He notes that in the modern church there is a suppression of ecstatic experience, with no discussion of their significance for life and faith. This causes modern Christians to possess an untapped "convictional unconscious" which is due to theological repression. Loder notes that such experiences need to be placed in a broader context, to understand how they work, for in modern Christianity the analysis of ecstatic events tends to end at their legitimation.

Modern religions tend to have goals that are not ecstatic, and as such they are uncomfortable with people who have such experiences. As Russell Shorto notes in his book *Saints and Madmen: Psychiatry Open Its Doors to Religion*, religions dread ecstatic states:

> Religions have been built up around it in order to harness it, to make the ecstatic flash that comes with breaking through the I- it boundary useful,

> but as religions become institutions they inevitably come to downplay and belittle and mock the mystical experience. It is, after all, so very different from them. They are of the social fabric: rational, sober, with duties that extend from guiding the upbringing of children to ensuring that dead bodies are thoughtfully disposed of. Religion is heavy with responsibility. Mysticism is none of those things...[59]

On the issue of analysis of ecstatic states and the problems of theological repression, we may examine the recent events in the town of Medjugorje.

Theology and Pathologies, the Case of Medjugorje

An interesting modern example of pathologizing of ecstatic experience has occurred in Medjugorje, in response to the apparitions of the Virgin Mary which were seen by six children. These visions began in 1981, and they continued on a daily basis. The children described their experiences as communications from the Virgin Mary, appearing as the Queen of Peace. This is perhaps ironic, as her visions appeared in Yugoslavia the midst of some of the worst fighting and atrocities since World War II. It is perhaps the most studied complex of ecstatic events in history, or at least as Randall Sullivan notes, it generated "the most complete scientific study of an alleged supernatural event in the history of Catholicism."[60] Medjugorje, a town "between two hills" and Roman Catholic pilgrimage site in southern Bosnia-Herzegovina, has grown into one of the largest Marian devotional sites in the world, larger than Fatima, and with almost as many visitors as Lourdes (since 1981, over thirty million people have visited the site). As Medjugorje gained momentum as a holy site, the area around it descended into factional religious and ethnic warfare. Armed conflict began in the early 1990s and continued into the latter half of that decade, ending with the breakup of Yugoslavia.

Sullivan is a writer for *Rolling Stone*, and in his book *The Miracle Detective: An Investigation of Holy Visions*, he analyzes several different claims of miraculous and visionary events. He describes the major older objections to miraculous events—fraud, hysteria, demonic possession, and includes the seventeenth-century standards written by Pope Benedict XIV: if the ecstasy is followed by weariness, forgetfulness, sluggishness, or sadness, it is due to mental illness; if has indecent movement and contortions it is from the Devil; if there is tranquility, humility and cheerfulness it is divine ecstasy.[61] Sullivan examines the arguments traditionally given by the Promoter of the Faith (also known as the Devil's Advocate, a position eliminated from the Catholic Church in 1983) against candidates for sainthood, which often involves the analysis of ecstasies and miracles.

While Sullivan investigates a number of different ecstatic events in his book, the major focus is on Medjugorje. He spent several years interviewing informants there, and he gives a detailed description of events associated with the visions. The children who reported the visions were subjected to many stresses: They were taken by the police and thrown into jail, rushed to the hospital in ambulances, examined by pediatricians for drugs, and taken to the morgue to be frightened by the corpses. They were left with mentally ill people in the psychiatric wing of the hospital, and threatened with being locked in the psychiatric ward (as did happen with political prisoners in Sarajevo). They were threatened by the secret police, interviewed, and tape recorded. Most local priests were hostile and insisted the visions were the work of Satan, and told the children they would go to hell, while the police threatened them with jailing their parents.[62] Social workers were sent into find family problems, and the Communist League threatened their parents with loss of their jobs.

The local bishop claimed these were false visions and formed a hostile investigative group, to examine doctrine and morals, and prove that their experiences were fraudulent. A priest of this group repeatedly stabbed one of the children with a leather-working needle while she was in ecstatic prayer, and she continued praying while blood was running down (she did not react to the needle). The children were watched and repeatedly, questioned, and followed by police officers. One girl was taken to the police station every day for six months, threatened and ridiculed. The priest most sympathetic to them was imprisoned. The other children at their school mocked and taunted them relentlessly, and they were isolated from former friends.

By 1983, foreigners were coming to investigate the visions, and a major team was led by Dr. Henri Joyeaux from the University of Montpelier. They made video and audio tapes during the time where the children experienced thirty-five apparitions, testing them with electrocardiograms and electroencephalograms to eliminate the causes of the visions as dream, sleep, and epilepsy. The brainwaves of the children during contemplation showed up as consistent alpha waves, with no beta waves present.

When the research report was published in 1985, it noted that the state of the children appeared to be something like deep prayer. A specialist in ear, nose, and throat diseases with the team blared ninety decibels of engine noise in the children's ears during their visions, and they had no response to this. He described what appeared to be a "disconnection of the auditory pathways during the ecstasy" of each visionary.[63] An ophthalmologist noted the inhibited eyelid reflexes of the children when suddenly confronted with beams of bright light during prayer—they did not blink. The team also noted the synchronization of their eyeball movements, which measured less that a fifth of a second apart at both the beginning and the end of their apparitions. The ophthalmologist stated that "this was so far beyond the capacity of normal human functioning that no form of collusion or manipulation could account for it."[64]

Some of the medical teams were quite harsh to the visionaries. A team of doctors from Milan were led by Dr. Michal Sabatini, who created a machine to test degrees of pain, which he called an algometer. Despite causing great pain by means of severe pressure on sensitive parts of the children's bodies during ecstatic prayer, they showed no response to the pain or "complete analgesis" during neurological tests.[65] There were teams from Italy, Poland, Austria, the USA, and England, who were experts in psychology, "religious delirium," psychiatry, internal medicine, neurology, gynecology, psychophysiology, and hypnotherapy. The children were given a variety of tests: MMPI, EPI, MHQ, Rorschach, Tree test, Person test, Raven Matrixes, Hand test, and Valsecchi truth detection test. They were given simultaneous polygraphs and electrocardiograms. Though these did not demonstrate dishonesty or epilepsy, they did (unsurprisingly) show stress.[66] While psychiatrists were able to hypnotize the children, they could not produce any visions of Virgin Mary during the hypnotic trance.

The miraculous healings associated with pilgrimage to Medjugorge were examined by a new Bureau of Verification of Extraordinary Healings (created in 1985, and modeled on the Bureau Medicale at Lourdes). They claimed a higher standard of proof than the Vatican's Consulta Medica and analyzed over five hundred allegedly miraculous healings which occurred at Medjugorje (most healings were diseases of the nervous system, especially multiple sclerosis).[67] By 1990, there were over three thousand apparitions that had appeared to the children, as compared to twenty-four at Lourdes and Fatima combined.

It is clear how the pathologizing predisposition is followed in this case by medical, psychological, and theological authorities. We have everything from sophisticated psychological tests to stabbing the ecstatics while they were in prayer, in order to find faults or inconsistencies in their claims. While there is much that can be done to disprove the validity of their descriptions of the apparitions, there is little or nothing that can be considered as evidence of their truth. The focus of church investigators was on possible heresy and error expressed through the visions, and whether there was any evidence of physical or mental disease in the ecstatics. Observers have been struck by the indifference of the theologians to the religious events, and the caution and skepticism shown by the ecclesiastical hierarchy. Sullivan notes a fear in the church of looking irrational and gullible; false miracles could cause embarrassment, and perhaps hysteria, and it is safer to disprove them.

The Vatican has not yet made an official judgment on the Medjugorje visions, though a commission has accepted Medjugorje as a place of prayer, and in 2017, it was accepted as a place of pilgrimage. In 2010, Pope Benedict XVI began formal investigations of the apparitions.[68]

The tendency to pathologize the ecstatic events is also shown in publications by scholars who have examined the visions at Medjugorje through the lens of political science, sociology, and anthropology. The interest here is in the social pathologies surrounding the visionaries, reducing the causes of the visions to local conflicts. There are accusations of infighting between church groups, clan ethnicities, misuse of donations, money laundering through pilgrimages, advantages from religious tourism for commercial establishments based on the visions, and excuses for warfare. There are questions of who appropriates the miracle, and to what end. Sociologist Raymond Eve writes of the figure of Mary as motivating warfare in his article "Politicizing the Virgin Mary: The Instance of the Madonna of Medjugorje." He calls the apparitions a result of pseudoscience arising out of cultural and religious forces, and argues that below the peace and love of the Madonna's words we have superstition, claims of the paranormal, and a potential Catholic nationalist revitalization movement. This is followed by an appropriation of the miracle by the Catholic Church, competition between Catholic religious orders, and children in a war zone with a desire for nurturing and stability. Instead of being a deity's desire for peace, the visions were symbolic weapons of war, creating worse tensions in the region.[69]

Anthropologist Lehel Peti describes the ways that social fears are reflected in ecstatic experiences in his article "Collective Visions in the Moldavian Csango Villages." He discusses the articulation of crises in visions, which give certainty about the religion, and security in uncertain situations. He calls Medjugorje "an unholy alliance of gullibility and politics" with its insanity and hysteria. He describes it as a situation which demonstrates the ways that religion disguises social problems.[70]

Anthropologists also examined the involvement of blood feuds between Croat clans, as well as between the warring factions of Croatian Catholics, Eastern Orthodox Serbians, and Muslims. Political anthropologist Mart Bax writes of war refugees mixed in with pilgrims, who went to Medjugorje as sanctuary from the war in Yugoslavia. However, it was not a reliable sanctuary—there were instances of personal retribution at night, with about eighty people killed in 1991.[71] He also notes in his book *Medjugorje: Religion, Politics and Violence in Rural Bosnia*, that there are multiple levels of violence, "small wars" within greater wars, and that these "lower levels of social integration" are unpredictable and erratic. The ecstatic states of the children motivated religious violence, the creation of "imaginary kingdoms," religious attachment to the pilgrimage site, and justification for a tribal network based on blood feuds and vengeance.

This violence over perceived past grievances has been going on since World War II, during the occupation of the nation by Nazi Germany. The Croats were on the side of Germany, who had established a fascist state, against the Serbians, who sided with the Communist partisans and had their own concentration camps. Both were angry at the Muslims after four hundred years of oppression under Turkish rule. In 1945, Yugoslavia became Communist under Tito, and religion was banned from the classrooms. However, the long history of abuses and atrocities were suppressed "behind the official communist rhetoric."[72] It flared up with the Bosnian war and Milosovic's language of religious crusade, encouraging people to act out their nationalist ambitions in violent ways. Past violence became an excuse for war between all of the major factions, as well as blood feuds between Croat clans and infighting between religious groups.

Ethicist Joseph Wiinikka-Lydon writes on the conflation of transcendence and nationhood in his article "The Ambivalence of Medjugorje: the Dynamics of Violence, Peace and Nationalism at a Catholic Pilgrimage Site during the Bosnian War." Ethnic violence used religious claims as a cover, with the ecstatic experiences interpreted as a justification for war. As the Virgin Mary appearing to Croatian Catholic children, this meant that she supported their nationalist cause.[73] Virgin Mary was understood as the protector and queen of Croat culture, and Croatian politicians would defend her against the Serbs and Muslims. As Mary only spoke against war generally and not against specific atrocities, they took that to mean that she overlooked atrocities. They chose to focus on victimization from many centuries ago, and simply ignored what Mary had to say about respect for other religions.

Lydon mentions Vamik Volkan's concept of "chosen trauma," in which the identity of a large group (ethnic, national, or religious) is defined as the subjective experience of thousands or millions of people who are linked by a persistent sense of sameness, though they may share other local qualities. The major obligation that members of the large group share is to maintain and support their group identity. The term "chosen trauma" refers to the shared mental representation of a massive trauma that the group's ancestors had suffered in the past at the hand of an enemy. When a large group feels endangered, its chosen trauma is reactivated in order to support the group's threatened identity. It may then become the basis of war or group violence against that enemy.[74] In such circumstances, the tendency is for the large group to identify itself as victimized and to ignore its own aggressive actions.

As we can see, both at Medjugorje and in the broader realms of academia and Theology, ecstatic states have been condemned from many perspectives. Indeed, they are a sort of Rorschach test, depending on the intellectual identity of the observer. Can they be negative from every possible angle? Can one set of phenomena be simultaneously depraved, superstitious, inferior, leading to homicide and suicide, compensatory, weak, hormonally imbalanced, undernourished, forbidden, and schizophrenic? And in this section claiming gullibility, hysteria, epilepsy, hypnotic states, Satanic action, and religious delirium, we are just looking at Christian ecstasy.

We categorize and pathologize differences, which become threatening, unwanted, unacceptable, and politicized. It is difficult for ecstatic states to be the result of so many negative causes at once. Most of these approaches are not proofs, in any experimental sense, they are instead interpretations and claims combined with personal values. However, one thing that all of these theories agree on is that ecstatic states do not belong in the modern West. Those states that persist must change and take on other forms. In the next two chapters, we shall look at some ways that states of religious ecstasy have been adapted to this situation, through finding ecstatic joy in violence and destruction, and interpreting otherwise secular events in a spiritual fashion.

Notes

1. This term became popular after its use in Donald Wiebe's 1984 article, "The Failure of Nerve in the Academic Study of Religion," in *Studies in Religion*, #13, pp. 401–422. It was used in theory and method debates, often to attack writers who did not use theological language or assumptions, but still seemed somehow too sympathetic to religious experience.
2. See Megillah 14a.
3. Jacob Neusner, *Theological Dictionary of Rabbinic Judaism: Principal Theological Categories* (Lanham: University Press of America, 2005), p. 139, in chapter 15 on Prophecy. The commentary is on Mishnah tractate Sotah 9: 12A.
4. Ibid., p. 139.
5. Ibid., p. 141.
6. J. Lindblom, *Prophecy in Ancient Israel* (Philadelphia: Fortress Press, 1961), pp. 4 ff.
7. Abraham Heschel, *The Prophets* (New York: Harper & Row, 1962), p. 405 (185, paperback).
8. Ibid., Pesikta Rabbati 35.
9. Ibid., Sefer Hasidim; also R. Elijah of Vilna, commentary on Seder Olam Rabbah 30.
10. For a longer discussion of the issue, see L. Stephen Cook, *On the Question of the "Cessation of Prophecy" in Ancient Judaism*. Text and Studies in *Ancient Judaism*, #145 (Tubingen, Germany: Mohr Siebeck, 2011).
11. See the *New York Times* article by Rukmini Callimachi, "To the World They Are Muslims, to ISIS Sufis Are Heretics," https://www.nytimes.com/2017/11/25/world/middleeast/sufi-muslims-isis-sinai.html.
12. From St. Augustine, "On the True Religion," cited in William De Arteaga, *Quenching the Spirit: Examining Centuries of Opposition to the Moving of the Holy Spirit* (Lake Mary: Creation House, 1992), p. 68.
13. De Arteaga, ibid., pp. 79–82.
14. Albert Farges, *Mystical Phenomena Compared with Their Human and Diabolical Counterfeits: A Treatise on Mystical Theology by Agreement with the Principles of St. Teresa Set Forth by the Carmelite Congress of 1923 at Madrid* (New York: Benziger Brothers, 1926). Farges founded the University Seminary at Angers in 1898, and lectured there on ascetic and mystical theology.

15. Ibid., p. 1.
16. Ibid., p. 9.
17. Ibid., p. 14.
18. Ibid., p. 15.
19. Ibid., p. 55.
20. Ibid., p. 155.
21. Ibid., p. 168.
22. Ibid., p. 201.
23. Ibid., p. 202.
24. Ibid., p. 324.
25. Ibid., p. 335.
26. Ibid., p. 356.
27. Ibid., pp. 356–357.
28. Ibid., p. 377.
29. Ibid., p. 444.
30. Ibid.
31. Ibid., p. 462, citing Dr. Lelut.
32. Ibid., p. 463, citing Dr. Jouffroy.
33. Ibid., p. 494.
34. Ibid., p. 584.
35. R. A. Knox, *Enthusiasm: A Chapter in the History of Religion, with Special Reference to the XVII and XVIII Centuries* (Oxford: Clarendon Press, 1950), p. 2.
36. Ibid., p. 2.
37. Ibid., p. 4.
38. Ibid., p. 6.
39. Ibid., p. 21.
40. Ibid., pp. 521, 523.
41. Ibid., p. 581.
42. Ibid., p. 585.
43. Ibid., p. 20.
44. Ibid.
45. Ibid., p. 43.
46. Fr. Benedict J. Groeschel, *A Still, Small Voice: A Practical Guide on Reported Revelations* (San Francisco: Ignatius Press, 1993).
47. Ibid., p. 24.
48. Ibid., p. 41.
49. Ibid., p. 85.
50. We may note that this allows for beginners to practice exorcism, which in the Catholic Church requires expertise.
51. Doris M. Wagner, *How to Cast Out Demons: A Beginner's Guide* (Colorado Springs: Wagner Institute for Practical Ministry, 1999), p. 47.

52. Ibid., p. 82. She notes that pornography is a special problem of Christian ministers.
53. Ibid., p. 126.
54. Ibid., p. 56.
55. Ibid., p. 83.
56. Ibid., p. 93.
57. This is cited from Noel and Phyl Gibson's book *Evicting Demonic Intruders*, in Appendix 4, ibid., p. 3.
58. James Loder, *The Transforming Moment* (Colorado Springs: Helmers and Howards, 1989), p. 18.
59. Shorto, ibid., pp. 198–199.
60. Randall Sullivan, *The Miracle Detective: An Investigation of Holy Visions* (New York: Atlantic Monthly Press, 2004), p. 201.
61. Ibid., p. 35.
62. Ibid., pp. 92–97.
63. Ibid., p. 202.
64. Ibid., p. 203.
65. Ibid., p. 204.
66. Ibid., p. 386.
67. Ibid., p. 215.
68. See online at http://www.usatoday.com/news/religion/2010-03-17-vatican-virgin-apparition_N.htm.
69. Raymond Eve, "Politicizing the Virgin Mary: The Instance of the Madonna of Medjugorje," online at http://www.csicop.org/si/show/politicizing_the_virgin_mary_the_instance_of_the_madonna_of_medjugorje/.
70. Lehel Peti, "Collective Visions in the Moldavian Csango Villages," in *Acta Ethnographica Hungarica*, vol. 54, #2 (Budapest: Akademiai Kiado Press, December 2009), pp. 297–308.
71. Mart Bax, "Barbarization in a Bosnian Pilgrimage Center," in *Neighbors at War: Anthropological Perspectives on Yugoslav Ethnicity, Culture and History*, ed. Joel Halperin and Devid Kideckel (University Park, PA: Penn State University Press, 2000), p. 201.
72. Mart Bax, *Medjugorje: Reilgion, Politics and Violence in Rural Bosnia* (Amsterdam: VU University Press, 1997), p. 126.
73. Joseph Wiinikka-Lydon, "The Ambivalence of Medjugorge: The Dynamics of Violence, Peace and Nationalism at a Catholic Pilgrimage Site During the Bosnian War," *Journal of Religion and Society* 12 (2010), online at http://moses.creighton.edu/JRS/toc/current.html.
74. Vamik D. Volkan, "Transgenerational Transmissions and Chosen Traumas: An Aspect of Large-Group Identity," *Group Analysis* 34, no. 1 (March 2001): 79–97.

CHAPTER 5

Destructive Ecstasies: Wargasm and the Joy of Violence

The documentary film *Fahrenheit 9/11* has a scene of a soldier in a heavily armored tank, talking about the "rush" he gets from killing Iraqis while listening to hard, violent rock music. It is a particularly clear example of what is one of the most widely known forms of ecstasy, the intense joy which arises from destruction. Such violence can come from war, from crime, and from many forms of domination. We see this ecstatic joy in Robert Duvall's character Lt. Colonel Kilgore in *Apocalypse Now*, riding into battle with Wagner's music blaring, loving the smell of napalm in the morning. Such destructive ecstasy is sometimes called "the thrill of the kill," the intense state that arises from violence. As General George S. Patton said observing the scarred French battlefield, "Could anything be more magnificent? Compared to war, all other forms of human endeavor shrink to insignificance. God, how I love it!"[1]

The destructive ecstasy that we see in violence and war can be useful and sometimes justified—it motivates people to fight and defend themselves, to protect others and kill enemies that threaten them. But for peaceful societies, the violent thrills that come from killing, crime, terrorism, gang violence, and raiding can become problems that drain large amounts of public money for police, courts, jails, and armies. The emotional intensity found in proximity to death and destruction may be secular, or it may be understood as a dark sort of religious ecstasy. Violent acts may be a way of giving transformative intensity and meaning to a dull life that lacks them otherwise, or there may be a vengeful God who inspires the struggle that fills both supporters and enemies with horror,

J. McDaniel, *Lost Ecstasy*, Interdisciplinary Approaches to the Study of Mysticism, https://doi.org/10.1007/978-3-319-92771-8_5

fear, disgust, and hatred. But it can also be a form of transgression, not only the secular fascination with the demonic, but a transformation of the supernatural and transcendent into the superstitious and dark. As Victoria Nelson has described the situation, it is a displaced religious impulse:

> Our culture's post-Reformation, post-Enlightenment prohibition on the supernatural and exclusion of a transcendent… has created the ontological equivalent of a perversion caused by repression. Lacking an allowable connection with the transcendent, we have substituted an obsessive, unconscious focus on the negative dimension of the denied experience.[2]

She notes that our society "carries the burden of the unacknowledged and unacceptable divine," as the transcendental has shrunk from awesome to awful, as Christianity turned pagan gods into Devils, and Hellenistic *daimones* into demons.[3] This is the modern Zeitgeist, where mystery becomes danger, and the miraculous becomes dark and dangerous. She links this with the broader Western rejection of Platonism in favor of Aristotelianism, where "the entire discredited worldview of Platonism eventually came to be linked with the underworld and the demonic grotesque."[4] Experiences banished from mainstream intellectual discourse turn up in other ways. This is Freud's "return of the repressed," except that here it is religious ecstasy which has been suppressed by the culture.

Observing the positive aspects of ecstasy is indeed difficult, for as we have seen in Chapter 4, there is a wide swath of academics and theologians suppressing the value of ecstatic experience. This chapter includes some of the things that they fear. Here, we shall explore two varieties of the dark return of the transcendent, as destructive ecstasy shown through violence and through transgression.

Ecstatic Violence

As Lars Bang Larsen notes in his article "When the Light Falls: Notes on Ecstasy and Corruption, "There is ambivalence in ecstasy, a dark undertow… ecstasy can be absolute freedom or a dance of death."[5] Following Immanuel Kant, he defines ecstasy as the sublime, a meeting of human consciousness with formless and immeasurable nature. It is not only like a starry night or a vast desert, but it includes the violence of destruction, found in hurricanes and volcanoes. Encounter with the sublime is the human response to the infinite, which is chaos beyond human comprehension. When one is carried away, reason is lost. As Larsen notes, "This is why systems of belief and knowledge based on reason and truth have official policies against ecstasy."[6]

This approach is based on the belief that if one seeks to transcend reason, what will be found is chaos and darkness. The irrational is decadent, fallen, making self-control impossible. Reason protects the person from the deeper nature of the self, which is a dark and dangerous thing. This deeper nature of the self may indeed be passionate and fallen in Christianity, to be avoided by reason. However, in Hindu Vedanta, the deeper self is ideal and virtuous, the *atman*, one's divine nature. One's view of ecstasy may well be based on one's cultural assumptions about human nature.

However, ecstatic states may be open to both interpretations. When the positive aspects of ecstasy are excluded, people turn to more negative ones. As the psychologist Mihaly Csikszentmihalyi writes in his book *Flow: The Psychology of Optimal Experience*, people need the state of flow, in which there is exhilaration and enjoyment, when body and mind are stretched to the limit and overcoming challenges. The optimal state is self-transcendence, sometimes with religious meaning. But when such states are limited, and people cannot find a state of flow in a positive way, they turn to the intense enjoyment of danger and random violence. Csikszentmihalyi includes Roman gladiators, Spanish killing of bulls, American boxing, fighting on the front lines of battle in war, and criminal behavior as places where the dark side of flow can be found.[7] It is not an opposition but a continuum, which may have different responses to the same events. He gives as an example sexuality, which can be "painful, revolting, frightening, neutral, pleasant, pleasurable, enjoyable or ecstatic" depending on its interpretation. It can be positive, a sport with physical skills, a meaningless ritual, or addiction, depending on individual understanding. The lack of flow, leaving activities that lack meaning, can make life impoverished or in some cases evoke violence as its antidote.

As William F. May writes in his article, "Terrorism as Strategy and Ecstasy," violence is a way of surpassing the limits of normal life, of transcending an ordinary identity. It is like an alternative kind of religious behavior, with sacramental ritual and leadership:

> [Ecstasy] literally means "to stand outside oneself," that is, to stand outside the limits of ordinary consciousness or to stand free of the restraints and limits of everyday behavior. Terrorism- whether of the established

> regime or the revolutionary left- is characterized by this ecstatic element… special uniforms, masks, sunglasses in Haiti, white robes and hoods for the Ku Klux Klan- all these devices emphasize the distinction between… the everyday world and the consecrated activities of those who… justify and apotheosize a dreadful violence.[8]

The term "wargasm" was used by Robin Morgan in her book *The Demon Lover: The Roots of Terrorism*, but its original use was by the radical group Weathermen, declaring their opposition to the United States government in 1969. For Morgan, a feminist writer, ecstatic violence is associated with both sexuality and death, and is a function of the patriarchal focus on power. Violence is hypnotic and attractive, as personified in the form of the Demon Lover:

> [He] is the ultimate sexual idol of a male-centered cultural tradition that stretches from pre-Biblical times to the present: he is the logical extension of the patriarchal hero/martyr. He is the Demon Lover, and society is (secretly or openly) fascinated by him… his intensity reeks of glamour… He is sexy because he is deadly, he excites with the thrill of fear.[9]

Both religious and secular ideologies glorify death and male leadership. Morgan quotes Kiramat Ullah, a fighter in Pakistan, on the glories of war:

> When we start fighting and bullets are flying and we are firing at the enemy and they are crying out and in trouble, and when some of my men are being injured and becoming martyrs- that is the peak![10]

She also cites the combat-training songs for the Rangers, Green Berets, and Delta Force members at Fort Bragg, North Carolina: "Glo-ry, Glo-ry, what a hell of a way to die!" As Randall Terry, founder of Operation Rescue, states in publicly inspiring his followers: "I want you to let a wave of intolerance, of hatred, wash over you…"[11] For these groups, it is fighting and death that makes life worth living; it is death that makes one alive. She calls it the "Sadean high." Morgan's fascination is the attraction of violence, power, and death. As she describes her Demon Lover figure, he is a pirate, an outlaw, a daredevil, a swashbuckler, and highwayman, seeking the ecstasy of death:

> He glares out from the reviewing stands, where the passing troops salute him... He straps a hundred pounds of weaponry to his body, larger than life on the film screen... He drives the fastest cars and wears the most opaque sunglasses... whatever he dons becomes a uniform. He is a living weapon.[12]... Recognizing only the redeeming ecstasy of a tragic death, the hero already lives as a dead man. As a dead man he is fearless, because as a dead man he is unconquerable by any life force.[13]

She places herself in his position:

> *But this*- the hands shaking, the throat dry, the heart pounding, the brain in a blur of excitation, the body poised, exhilarated, the risk of being swept into obliteration, the aphrodisiac in demanding power- *surely this surpasses whatever they mean by sexual joy.*[14]

Such a patriarchal "politics of ecstatic death" suffuses both nationalism and religion, appearing in sermons, political speeches, and the rush to war. For Morgan, it is a sensualization of cruelty and death, a lust for annihilation that opposes a more feminist ecstasy, in which the person stands inside of the self, rather than outside of it, and has compassion for the victims of violence and war. But the "high" of soldiers comes through ideology, from faith in leaders and traditions.

She notes that violence can come from secular and nationalist sources, as in terrorist groups like the Algerian FLN, the Italian Red Brigade, and the Tupamaros. It can also come from groups that have religious associations, such as the Muslim Hamas, Hezbollah, and Islamic Jihad groups, the Japanese Aum Shinrikyo, the Christian Army of God and The Lord's Resistance Army. In such cases, whether there is religious or secular ideology, it is rare that the individual members report their own current ecstatic religious experiences; more frequently, we see great faith and love toward idealized leaders, and a distant anticipated ecstatic goal that requires violence and sacrifice. That goal will hold the joy and beauty that this life lacks, and violence is justified, for it leads toward the ultimate goal.

One of the most well-known articles on the ecstasy of war was written by William Broyles, Jr in the November 1984 issue of Esquire Magazine. It was called "Why Men Love War," and it described the states of intense joy and excitement that he saw and felt during his years as a soldier in the Vietnam War. He was struck by the ways that people did not discuss this when they came back from the war, and wanted people to know about it. There were many unspoken attractions of war.

> The love of war stems from the union, deep in the core of our being between sex and destruction, beauty and horror, love and death. War may be the only way in which most men touch the mythic domains in our soul. It is, for men, at some terrible level, the closest thing to what childbirth is for women: the initiation into the power of life and death. It is like lifting off the corner of the universe and looking at what's underneath. To see war is to see into the dark heart of things, that no-man's-land between life and death, or even beyond.[15]

War brings about ecstatic states, because of its life or death intensity. He notes, "War stops time, intensifies experience to the point of a terrible ecstasy." It is a corrupting influence—as with animals, the taste of blood changes people:

> A lieutenant colonel I knew, a true intellectual, was put in charge of civil affairs, the work we did helping the Vietnamese grow rice and otherwise improve their lives. He was a sensitive man who kept a journal and seemed far better equipped for winning hearts and minds than for combat command. But he got one, and I remember flying out to visit his fire base the night after it had been attacked by an NVA sapper unit. Most of the combat troops had been out on an operation, so this colonel mustered a motley crew of clerks and cooks and drove the sappers off, chasing them across tile rice paddies and killing dozens of these elite enemy troops by the light of flares. That morning, as they were surveying what they had done and loading the dead NVA–all naked and covered with grease and mud so they could penetrate the barbed wire–on mechanical mules like so much garbage, there was a look of beatific contentment on the colonel's face that I had not seen except in charismatic churches. It was the look of a person transported into ecstasy.
>
> And I–what did I do, confronted with this beastly scene? I smiled back, as filled with bliss as he was. That was another of the times I stood on the edge of my humanity, looked into the pit, and loved what I saw there. I had surrendered to an aesthetic that was divorced from that crucial quality of empathy that lets us feel the sufferings of others. And I saw a terrible beauty there. War is not simply the spirit of ugliness, although it is certainly that, the Devil's work. But to give the Devil his due, it is also an affair of great and seductive beauty.[16]

Like Robin Morgan, Broyles discusses the sexual charge of war and the ways that death and sexuality seem to merge:

> Many men loved napalm, loved its silent power, the way it could make tree lines or houses explode as if by spontaneous combustion. But I always thought napalm was greatly overrated, unless you enjoy watching tires burn. I preferred white phosphorus, which exploded with a fulsome elegance, wreathing its target in intense and billowing white smoke, throwing out glowing red comets trailing brilliant white plumes. I loved it more–not less –because of its function: to destroy, to kill. The seduction of War is in its offering such intense beauty–divorced from all civilized values, but beauty still… Most men who have been to war, and most women who have been around it, remember that never in their lives did they have so heightened a sexuality. War is, in short, a turn-on. War cloaks men in a coat that conceals the limits and inadequacies of their separate natures. It gives them an aura, a collective power, an almost animal force.[17]

This description in some ways parallels that of Robin Morgan, with the intensity of violence as both sexual and destructive. It is an ecstasy she finds more suited to men.

When war is supported by the culture, there are often rituals and myths which give ultimate meaning to the violence, transforming cruel and sadistic acts into moral actions, expressions of virtue which turn warriors into heroes and saviors. The destructive ecstasy of violence and bloodshed is justified and redefined as positive and socially beneficial by brotherhood and the value of one's own group. The fascination and revulsion in the killing of enemies is balanced by the intense closeness between fellow soldiers. Sebastian Junger talks about this aspect of war:

> War is hell, as the saying goes–but it isn't only that. It's a lot of other things, too–most of them delivered in forms that are way more pure and intense than what is available back home. The undeniable hellishness of war forces men to bond in ways that aren't necessary–or even possible– in civilian society. The closest thing to it might be the parent-child bond, but that is not reciprocal. Children are generally not prepared to die for their parents, whereas the men in a platoon of combat infantry for the most part are prepared to do that for each other.[18]

War justifies both love of allies and ecstatic violence toward enemies, lending both heroism and morality. This sort of moral justification is a major problem with modern terrorist groups, who claim that killing enemy civilians without warning is morally justified by a future goal. Violence can be justified by love or by hatred.

Violent and destructive ecstasy can appear in other contexts than war. The psychoanalyst Michael Eigen writes of the prominence of destructive ecstasy in his patients:

> There are destructive as well as creative ecstasies, ecstasies of war and injury, brutal ecstasies.... Ecstasies of pain and trauma confuse souls and mind. In light of miseries wrought by ecstatic destruction, no critique of ecstasy can be too strong. History is filled with stunning abuse between groups and persons fueled by ecstatic processes that sour, take cruel ideological turns, pit being against being. Ecstasy and its twin, ecstasy envy, readily meld into a righteous rage that can be all too calculating.[19]

Eigen can also see the more positive, primal side of ecstasy which is focused in the body and is neither positive nor negative:

> Ecstasy is the heart's center. The heart of life. It is not reserved for the soul's union with God, although that is where it starts and ends. It pervades the body, the inside and outside of the skin, the pulsing of organs. It is in the senses... and blood.[20]

However, he recognizes that the ecstasy associated with the blood can be very dark:

> Blood ecstasies can be terrible. Not just ecstasies of sex, but ecstasies of murder, ecstasies of fear and rage. There are patients who must cut themselves, see and smear and taste their blood, not only to feel real but to feel ecstatic. There are individuals who must cut others to tap a stream of ecstasy... Death-life feed each other, soul of the hunt. Twist the sense of power and you have the frenzy of Nazi calculation, chills of exaltation and stupor, extermination ecstasies.[21]

As Eigen notes, a part of this search for ecstasy can be sexual, from a sort of sexual desire that is inadequate, blocked, or armored. He notes, "the schizophrenic and Nazi share orgasmic inadequacy, they seek alternative orgasm: madness and murder."[22] But the ecstasy which is wholeness, integrating all parts of the self, requires more than that. Rather than "sinking" or "merging" or "being filled" with God, people seek to tear away what is imperfect and impure, literally cutting their flesh and tearing away their ideas and emotions. Eigen quotes Plotinus on how to get to God through ecstasy—"cut away everything"—and some people take this sort of approach literally:

> Their blades want something that can't be touched by metal. Do they feel something ineffable for a moment when they see blood? The soul is in the blood? Or perhaps they angrily insist body is a gateway to ecstasy, the bloody body. No body, no ecstasy.[23]

This is how ecstasy is sought in a materialist culture—in and through the body. While some people can find destructive ecstasy on their own, attacking their own bodies by self-mutilation, most people require enemies, real or imagined. People need an Other to fight, to create a true intensity of hatred, and to focus the hatred and anxiety in life. The enemy becomes the scapegoat, and the ecstasy of destruction is loosing anger upon it.

Violent ecstasy may be motivated by both secular and religious ends. It appears in secular contexts which do not support any transcendent form of ecstasy, which the culture promotes with a variety of extremes—extreme sports, extreme combat, and extreme terror at horror films.[24] Intense experience becomes either illegal, especially through the drug wars, or dangerous, in survival shows and violent competitions, in random sexuality, addiction, and terrorism. Simulations of dangerous situations become popular entertainment, found on television crime shows, reality survival shows, and in war films. As Murray Pomerance notes in his edited volume *Bad: Infamy, Darkness, Evil and Slime on Screen*, violence and corruption "sell big." People are attracted by screen representations of negativity: evil, monstrosity, corruption, ugliness, villainy, and darkness, including killers, vamps, cold-blooded megalomaniacs, and demons.[25] His book has chapters on Nazis, violent pornography, and "bad goodness." Violence and corruption on screen can evoke emotions like terror, disgust, and rage in the audience. Homicidal ecstasy, which emphasizes the lure of the forbidden, and criminal behavior are made attractive.

We may also see a form of destructive ecstasy in the popular fascination with films depicting monsters, especially vampires. Without positive spiritual figures to represent immortality, we move to the demonic. The vampire is an immortal being who lives intensely and passionately, a cultural rebel, and a tragic figure whose lovers are victims who die while he lives on. They have cosmic longings, unlike dull humans, and are "born to the blood," feeling the intense draw of the night. They are different from the herd of humanity, both more sensitive and more callous. The older image of Dracula has become more popular through Anne Rice's novels, in which vampires roam history as predators, taking what they choose. They are powerful and instinctual, living in both natural and supernatural worlds. The endless life of vampires does not have immortality as an ideal heaven, but rather an infinity of darkness and violence. When there is no positive heaven, a dark underworld can substitute as immortal life.

Their ecstatic states are supernatural and destructive, taking both blood and life from their victims. They are instinctual, sensual, and brilliant but corrupted by what they must do to live. An example of vampire ecstasy comes from Rice's *Interview with a Vampire*, in which the vampire Louis de Pointe du Lac is an aristocrat who becomes a vampire, attracted to the "pulsing veins" of his male and female victims:

> I sank my teeth into his skin, my body rigid, that hard sex driving against me, and I lifted him in passion off the floor. Wave after wave of his beating heart passed into me as, weightless, I rocked with him, devouring him, his ecstasy, his conscious pleasure.[26]

In vampire literature, ecstasy is associated with both drawing blood and having it taken. For Anne Rice, vampires are passionate and tragic figures, caught in destructive passions. The vampire ecstasy of attacking victims and drawing their blood can be associated with aggression, rage, hunger, and sexuality. As David and Ellen Ramsdale note, rage and sexual orgasm are similar physiologically. There is rushing blood, facial contortions, expressive sounds, and release of stored energy. They note, "Getting angry and expressing it very dramatically may be a socially acceptable way of having an orgasm-type experience in public, a sort of "angergasm."[27] Part of the appeal of vampires is their public expression of private, forbidden passions, and the linking of violence and sexuality.

Destructive ecstasy can also be simulated by watching violent films, as in the recent genre of films that include detailed scenes of torture. Rage becomes mixed with sexuality, as people become enthusiastic over prolonged torture scenes in films. As Jenny McCartney notes in her article, "The rise of 'torture porn'," recent horror films,

> ...openly beckon the audience to relish voyeuristically the victim's prolonged agony, and often to enter into a form of dubious complicity with the 'sicko' who wields control. They recognize few boundaries to the degradation they will heap upon their protagonists for the delectation of their audience.[28]

Thus, we can see different styles of destructive ecstasy. One is the ecstasy of sheer destruction, found in battle frenzy, the joy in killing, the glory of revenge, the adrenaline rush of war. Another is an ecstasy of deliberate sin, of transgression, in which the goal is not necessarily death, but rather pain and harm, the violation of boundaries, and artistic originality found through the recognition of human misery.

The Ecstasy of Transgression

One well-known writer and literary critic who justifies this sort of violation, which he calls "transgressive ecstasy," is Georges Bataille. Bataille describes human life as ordinary, conformist, robotic, confined by the past. How can one be original, think new thoughts, be a true artist or writer or thinker? Certainly, we cannot transcend the ordinary in a religious sense, as there is no God and there are no religious goals to achieve; people cannot reach any real sphere of transcendence. People cannot find God; they only go through levels of meaning upon meaning, a "hermeneutical schizophrenia" which searches for some God substitute. The only real potential for originality comes through what he calls "transgression:"

> To define transgression we must think of a threshold, or rather, a movement towards the threshold, towards the limit, where there is no longer interpretation. We must think of the self being pushed to its own limits... Each limit is revealed through transgression...And with each new understanding, an abyss is opened and a new fall takes place. But transgression offers the hope that the sacred can be recovered in the shadow of a dead god.[29]

Bataille's "paradoxical philosophy" arose from difficult and violent periods in his own life. He described the world as painful and horrifying, full of violent sexuality, sickness, and intoxication, causing the "collapse of being into night."[30] In his writing, literature is transgression, "the dark and unholy rendering of a sacrificial wound" which allows true and meaningful communication, rather than ordinary pseudo-communication. It reveals the "violent absence of the good,"[31] the death of God and the absence of any sacred dimension of human life. This is the cause of terrible suffering, but it is meaningful because it is the true human situation. Divine ecstasy is horror for Bataille; it is intensity which must exist without any sacred dimension. Transgressive joy is the surpassing of annihilation, the realization of nothingness, and thus the knowledge that everything is permitted and nothing can be ultimately limited.

God is limitation, "wing-clipping"' that which forbids. One's freedom without God is intensity, and only hell remains, as heaven is gone. This is the hell of human suffering, the meaning which remains for people without a transcendent realm. When Bataille writes "I have wanted and found ecstasy,"[32] this is an ecstasy of excess, surviving the loss of heaven. He describes it as the time "when the plenitude of horror and that of joy coincide."[33]

Because there is no ultimate good, the ecstatic must strive for an intensity of evil—found in violence, murder, blasphemy, excessive sexuality. He states, "I search only for the terror of evil"[34] the denial of the false and rotting ordinary world, and the seeking of intensity and originality. Only fear and eroticism are powerful enough to negate nature and the profane world, thus the "erotic attraction of limits and taboos." Transcendence becomes transgression, ecstasy becomes excess, and because there is no sacred aspect to life, artistic originality requires the intensity of the profane. Thus, disorder and madness are superior to ordinary order and rationality, and we escape from the ordinary and repetitive world by transgressive ecstasy. Transgression is what remains when the sacred is gone, intensity without divinity, a fascination with horror, madness, violence, eroticism, and death. Some people find that fascination in suicide, an "intensity of the profane." Talal Asad notes,

> For most witnesses, horror- a compound of pain and delight or (as Bataille put it) of ecstasy and unbearable pain- is generated by the unexpected image of a broken body, a shattered human identity. There are few things as shocking as a sudden suicide in one's presence. A suicide operation, in which many die and are wounded, extends that shock. A possible refuge from horrified helplessness in that case is righteous anger directed at the perpetrator of the deadly violence. … In the suicide bomber's act, perhaps what horrifies is not just dying and killing (or killing by dying) but the violent appearance of something that is normally disregarded in secular

modernity: the limitless pursuit of freedom, the illusion of an uncoerced interiority that can withstand the force of institutional disciplines.[35]

We also see such violent imagery in the writing of Foucault, who speaks of interpretations devouring themselves in a world without a sacred. People seek some ultimate end to interpretation, but without success, and become intimately tied to madness. For Foucault, intensity without the sacred brings madness, horror, and the grotesque. For instance, the images of the crucified Christ are horrific and severe, full of potentiality for horror and madness. There is an abundance of dark meaning, with no higher goal to which it points. This is the "decay of Gothic symbolism" which is dizzying and frightening when it is not able to point to a spiritual object.[36] We have shattering and dismemberment—transgression, but no transcendence.

Thus, the lack of religious ecstasy leads to transgression as the only true originality, the only possible extreme in a Nietzschean world where God is dead and heaven is inaccessible. Transgression is immensely accessible—one need only violate a moral rule or a social norm. Thus, we have direct transgression—violence in relationships, drunken or drug-fueled rages, and passionate addictions. We also have vicarious transgression, in dogfights and bullfights where people get to watch violence leading to death, in film dramas of revenge and rape and degradation and violation of taboos.

In his article "Genocide as Transgression," Dan Stone describes genocide in ecstatic terms. Following on Roger Caillios' anthropological analysis of "war as festival," and Dominick LaCapra's descriptions of joy in scapegoating and the "carnivalesque" aspects of the Holocaust, Stone writes that any act of genocide involves a heightening of community feeling, and "ecstatic sense of belonging" when people share a normally forbidden act of transgression in order to safeguard the community by killing its designated threatening group.[37]

In his book *Memory, History and the Extermination of the Jews of Europe*, Saul Friedlander writes of the "mystical Fuehrer-bond" which seduced Germany into supporting the Holocaust. He spoke of *rausch*, the intoxication or ecstasy of destruction which occurs at the breaking of a universal taboo, in this case mass murder. Such irrepressible exaltation substitutes for a more traditional religious ecstatic goal, for this is human transcendence through existential experience, in which ordinary men are required to transgress moral laws in the name of a new historical epoch.

National Socialism combined the thrill of perfect obedience with a violation of the ordinary order. As Himmler stated in his 1943 speech to the SS, the mass murder of the Holocaust was glorious, "modernity's final and highest expression." But he exhorted his listeners to be silent about these deeds, as the world would not understand them.[38] This *rausch* drew ordinary people into becoming extraordinary, so that they might be transformed into secular reflections of the religious hero.

Destructive Ecstasy and Religion

Most religious war is not ecstatic—the soldier fights in obedience to a god's commands, in order to please the god. People generally do not have visions or mystical union with the God on the battlefield; they are servants of the God and his generals. Warriors are rarely prophets. The end of time may be full of ecstatic joy, but the present is a time of work and struggle.

Yet there are destructive images in many religions. The gods themselves may express and support violence in myths of past and future, as seen in Shiva's dance of destruction at the end of the *yuga* in Hinduism, the destruction of the world in *The Revelation of John*, and Yahweh's Biblical command to kill all of the Amalekites, even their children and their animals (1 Samuel 15:3). It is shown in the jihads and crusades of religious followers, listening to songs intended to evoke an emotional response, a "battle ecstasy" which can motivate more violent fighting. Such songs may range from religious chants of the Shahada in Islam, songs of the Soldiers of the Cross, to the Battle Hymn of the Republic. Destructive ecstasy may take on extreme forms—the heroes of the Greek epic *The Iliad* were possessed by gods during battles, and the "battle-fury" (*riastradh)* of the Irish hero Cuchulainn required multiple baths for him to cool down (the steam from his body burst three vats of cold water). According to the *Tain Bo Chuailnge (The Cattle Raid of Cooley)*, Cuchulainn's actions in battle involved supernatural power:

> During this slaughter, a graphic description is given of the battle fury, or 'riastradh', that Cuchulainn experienced from his early childhood. The change that comes over him is not just one of spirit but also a major physical one. His whole body changes until he takes on the appearance of an invincible, primeval dragon, with distended head and snapping jaws. During the course of the battle he himself, just like the Sí, was completely

> invisible to the enemy. After this great slaughter, Ailill estimated that almost half the number of men they had set out with, approximately 54,000 warriors, had been slain along the way by Cuchulainn.[39]

Destructive ecstasy in religious war can arise from killing enemies, or merely humiliating and degrading them. There are times when religious fervor can merge with destructive ecstasy, and religious warriors can be capable of a cruelty which often violates their own ethics.

Earlier forms of destructive religious ecstasy were often associated with raiding cultures. Among the Vikings of Scandinavia, some of the most fierce and powerful warriors were called the berserkers. Before battle, they would get into the trance state of *berserkergang*, a state which allowed them to ignore pain and danger. They dressed themselves in bear or wolf skins, to show that their violence was like that of wild animals. They would enter into the battle frenzy by biting their shields and howling, and they were ferocious fighters while this battle ecstasy lasted. Some tales say the berserkers were magically immune to weapons, unable to be harmed by fire or iron. The berserker is closely associated in many respects with the God Odin, who was said to shape-shift into the form of a bird, fish, or wild animal. The berserker, too, was often said to change into animal form, or at least to assume the ferocious qualities of the wolf or bear. We have descriptions of berserkers in battle frenzy going back to ninth-century Norway, and theorists have speculated about whether battle ecstasy was motivated by shamanic trance, possession by animal spirits, or the use of amanita muscaria mushrooms. A good poetic description of battle ecstasy with its distortion in the perception of time and space comes from a recent fiction novel, *Merlin*, by Stephen R. Lawhead. Myrddin Emrys, or Merlin, is on the battlefield:

> The cry of outrage reached my ears as a mild and distant sound, for I had once again entered that uncanny state where the actions of others were as languorous and slow as those of men half-asleep. The flying, careening warhost became a massive, lumbering thing, heavy-footed and dull, without speed or quickness, overcome by a languid torpor. Once again, as in the battle at Maridunum, I became invincible, dealing death with every well-calculated blow, hewing down mighty warriors with effortless strokes, my movements perfect in their deadly grace. The clash of battle reached my ears like the sound of water washing a far-off shore. I moved with elegant precision, striking boldly and with vengeance, my sword a living thing… I carved a swath through the enemy ranks as if I was a scythe and

> they the corn standing for harvest… I wept with laughter until I tasted the tears in my mouth.[40]

In Celtic tradition, battle ecstasy was a major way to become a hero. All of the major Western monotheisms have heroic strands which emphasize the importance of war, from the ancient struggles of the Israelites with the Canaanites, to the Crusades, to the armies of the Mujahaddin. Destructive ecstasy tends to be found in dualistic religions, in which the world becomes an arena of divine forces fighting for power. It becomes changed in the religious context is when hatred of the enemy becomes more intense and valuable than love of the god.

Destructive action may not actually bring the person ecstatic visions of the god, but it may certainly demonstrate an extreme form of obedience to the god. There have long been bloodthirsty gods, whose rage is only appeased through attacks on sinners, infidels, and heathen. Sometimes this involves death, though torture may be sufficient. In ancient Greece and Rome, for instance, there was voluntary whipping at Lacedaemon before the altar of Diana on the Day of Flagellations, and Plutarch notes that boys were sometimes whipped to death before Diana the Orthian—and the one who lasted longest was called the Hero.[41] Gods were believed to be calmed by torture—Petronius writes that Encolpus was whipped by sailors to prevent a storm and to appease the deity of the ship.[42] There were also Flagratores, who allowed themselves to be whipped for money, to calm the anger of the gods, or to rid a place of impurity. Historically, there are gods who have required individual human sacrifices, but the modern holy wars tend to require group sacrifices—we see genocides, martyrdom operations, and spreading religious Truth by means of the gun, the bomb, and the sword. Such behavior has been encouraged by warlike interpretations of religious texts in the past, and to a lesser extent messianic literature (especially apocalyptic fiction) in the present. We can see this emphasis on violence in certain of the modern fundamentalisms.

Mainstream religion in the modern West tends to emphasize ethics and reason, and avoids ecstatic extremes. Yet we have the creation of radical fundamentalist wings which emphasize ecstasy in the future, but only if the members obey the authoritarian leaders and support their wars and hatred of outsiders. One fact stands out in modern religious ecstatic violence: it tends to follow a vision of God which appeared in the distant past, or will appear at the end of the world. Ecstasy in the present world is absent, or in its rare occurrence highly suspect. Heaven is not on earth, and it is modern fighting and violence which will bring the faithful to that noble end. The closest that we can find to religious ecstasy is a sort of possession trance, when the person becomes God's avenging arm, or perhaps speaks a message from a prophet or historical figure. But such possession states are rare in the modern world.

Most religious forms of destructive ecstasy today emphasize the glorious past or the even more glorious future. Mankind must return to the golden age when people knew their place and worshipped the right god—and this must happen by force if it is not happening naturally. Or they must bring about the Day of the Lord, the righteous end of time, when God will judge the good and evil, and punish the wicked. Many of the so-called religious fundamentalisms emphasize that death will bring the religious warrior to heaven, or that indeed people may never have to die, as the end of the world and the Rapture will come during their lifetimes.

Modern attitudes toward destructive ecstasy may be seen in the science-fiction stories written within the worldview of modern fundamentalisms. This "finding truth in fiction" is perhaps ironic, with the fundamentalist emphasis upon literal truth.

Millennial writing often makes use of violent imagery. As an example, we can look at a few Islamist apocalyptic writers. Muhammad Isa Saud wrote of the return of the messianic figure of the Mahdi and the proclamation of the universal Caliphate in the near future, which would involve a nuclear battle between the Muslim countries and the West. The writer Bashir Muhammad identified the USA with the city of Ad, destroyed by Allah in the Qur'an (Ad has nuclear weapons, skyscrapers, arrogance, and tolerance of homosexuality). He describes a climactic battle at the end of time, in which the Jewish Zionists and Christian Crusaders (and also the Masons) will fight the armies of Islam and the Mahdi. New York City and its Jewish population will be annihilated by an earthquake, the rest of the USA will be destroyed, and the whole world will convert to Islam.[43] Muslims who live an obedient and ascetic life in normal times will get the joy of fighting alongside the Mahdi, the savior, and participating in the ecstatic genocide of the infidels. The signs of the Mahdi's earthly coming are predicted to be fear, confusion, violent deaths, *fitnas* (sins creating chaos), wars, earthly tyrants, and according to some interpretations, the appearance of Dajjal, the one-eyed Dark Messiah. As one hadith goes (a commentary on Surah 3:19 in the Qur'an), there are signs of the end-times. Hazart Abu Musa Al Ashari narrates that the Holy Prophet said that the Mahdi will come at Doomsday (Qiyamah)....[44]

The apocalyptic war to be led by the Mahdi is a well-known Shia end-time scenario, similar to the millennial ideas of many Christians. There are many Islamic sites on the Internet which talk about joining with the Mahdi's future army. The ideal is the future martyr—generally male, who gains respect, celebrity, and money for his family. He can wear white in the streets, symbolizing that he is pledged to death. In the 1979 Iranian revolution, soldiers wore white burial clothing to show their eagerness to die, with crimson martyr scarves and small "keys to heaven" around their necks. They followed Khomeini's motto—"To kill and be killed is the highest duty of every Muslim." As a young Shia man said, "I want to die before my friends. They want to die before me. We want to see our God."[45]

While the ideal of martyrdom is most well-known in Shia of Islam, it has also been adopted by some Sunni groups. According to an al-Qaeda spokesman, "There are thousands of young men who look forward to death... The *goal* is martyrdom."[46] Attitudes toward martyrdom and jihad war vary in the Muslim world. As Yaakov Lappin notes in his article on international terrorism, suicide becomes ecstatic:

> While many Muslim scholars have ruled that acts of jihad such as suicide bombing are forbidden acts of self-destruction, jihadi websites quickly establish that suicide terrorism is a holy sacrificial act and surround it in a mist of glamor and mysticism. It, like similar acts, is a pinnacle moment of spiritual ecstasy... While most soldiers are taught to avoid injury to themselves, jihadi recruits are told that coming back unscathed from their mission would be an unmitigated failure. Their harm and demise is necessary to reach heaven. "Either way there is victory, whether he kills or is killed," the website assured.[47]

While suicide is forbidden in mainstream Islam, it can be understood as valuable in certain circumstances. Suicide may indeed lead to the "pinnacle moment of spiritual ecstasy," as Lappin phrases it, or it may be an act that will bring the person to hell, depending upon the interpretation of the situation. Thus, it becomes modern Muslim theologians who determine if an act is heroic or sinful. Even the sacrifice of children can have a terrible beauty, as Iran's current president, Mahmoud Ahmadinejad, notes in the 2011 film "Iranium." When the Ayatollah Khomeini sent children on foot to clear minefields during the Iran–Iraq war in 1980, they were blown up by the mines. In the film's footage, he tells his followers of their deaths, "No art is more beautiful, more divine and more everlasting, than the art of martyrdom."[48] Many children accepted such deliberate death, as a way to reunite with their parents who had already died in the war.

Joyful participation in terrorist acts may be understood as either secular ecstasy or religious ecstasy for participants. For instance, the 9/11 attackers have been understood as acting based on religious ideals by Daniel Benjamin and Steven Simon. In their book, *The Age of Sacred Terror*, they emphasize how religion may be adapted to incorporate violence.[49] On the September 11, 2010 television show Frontline, there was a special on "Faith and Doubt at Ground Zero." The commentary noted the "thirst for the absolute" on the part of the bombers, interpreting the suicides as a religious search for absolutes, and an avoidance of nothingness. The suicides were interpreted as an act of worship, in which the religious idea of sacrifice for the sake of the community is turned into a private jihad, which then loses the distinction between combatants and non-combatants, and forgets the rules of war. It is a danger of the charismatic side of religion, in which one can ignore the law in favor of a direct line to God. This justification of ecstatic death and vengeance allow violence as a religious act, and dedicating the violence to the God makes it acceptable. Destruction becomes a form of piety.

However, we should also note that some accusations of destructive ecstasy over the course of time have been fantasies or exaggerations on the part of observers, often for political reasons. Much of the film imagery of orgies, sacrificing virgins, black masses, and various ritual reversals in non-Christian religions are more cinematic imagination than they are history. Such exaggeration of the dangers of foreign beliefs are not new. The witch craze, the persecutions of heretics during the Middle Ages, the European and Russian pogroms, and various holy wars through the centuries have been based on exaggerations of the dangers of outsiders and their religions, the Enemy Other who wishes harm for the sake of evil. But there may be other, more commercial, agendas. Some years back, Mad Magazine had a satirical article which asked the question, "What is really living?" They gave as examples fighting bloodthirsty sharks, skiing off high cliffs, and doing all manner of dangerous activities. They noted at the end of the article that it had been sponsored by the Funeral Directors Association of America. In typical Mad

Magazine style, it showed how special interests can redirect one's understanding of meaning in life, so that proximity to death becomes the sign of a meaningful life.

There are also many Christian millennial writings, in which the imminent end of the world will mean eternal life for the saved and eternal damnation for the unsaved. Many Millennialists look eagerly toward the wars at the end of time, led by Christ against the Antichrist. Perhaps the most famous apocalyptic text, the *Book of Revelation* by John, begins with John's contact with an angel to learn about the future (Rev 1:1–2). The book was thus described by its author as the result of a direct religious vision. In recent years, writers interested in the Book of Revelations and other apocalyptic writings have begun a genre of apocalyptic Christian novels. The most popular apocalyptic fiction series has been the *Left Behind* novels, though other modern messianic Christian series have followed in its wake. These books describe the imagined situation of the Tribulation after the Rapture, and a major focus is the sufferings of people who would not convert to Christianity. These sufferings were described in graphic detail. As an example, in the novel *Apollyon: The Destroyer is Unleashed*, those who were "saved" generally showed little compassion toward the people who were not saved, people in such intense pain that they put their heads in ovens and jumped from tall buildings, trying unsuccessfully to kill themselves. Instead, they were satisfied that they were right, and the others were wrong, and deserved their suffering. Their response was to attempt to convert the sufferers to Christianity. In reading some of these books, the major religious emotion seemed to be a form of *schadenfreude*, the joy in another's suffering. Only Christians would have access to true religious ecstasy, both in the Rapture and in the coming millennial kingdom.

Destructive ecstasy in the Western religions tends to be motivated by one salient fact—there is little opportunity for more positive and transcendent forms of ecstasy. This is because, in the Western post-Enlightenment secular world, prior to death, there is no possible access to a transcendent realm. How then does one have a truly intense experience, especially of the overwhelming religious sort? If there is no option for positive ecstasy, destructive ecstasy would have to substitute. We have available the milder secular ecstasies for those who fit well into the culture, victory in sports, or falling in love, or appreciating music, or winning the lottery. But for those who do not fit in, and have other values, we see that violence often substitutes for transcendence, bringing a dark form of ecstasy in the joys of war and struggle.

While polite Western society frowns upon the joys of blood-lust, there are institutions which support it. Soldiers should enter gladly into battle, and boxers look forward to the ring. For the untrained and the less athletic, there are games which simulate violence, and are intended to cause intense reactions in their players. There are video games based on murder, theft, and rape, and these have become popular in school-age children. While fairy tales have often been deemed too violent for children in the elementary school curriculum, we have more vivid and direct violence in the games that the children play. Entrance into virtual worlds in the computer entertainment subculture allows intense interactive gaming in animated environments; and such games are sometimes called the "ecstatic technologies."[50]

In many video games, the player changes identities, becoming a warrior or a thief or a criminal. As a person changes identities when he becomes a member of the Lord's Army or a *mujahid*, so a child playing a video game looks out through another's eyes. He perceives in a different way, as he adopts various roles that require violent actions. Simulated violence allows children to act as simulated criminals. Video games themselves have many levels of violence within them. Some games turn children into warriors, others allow children to experience criminal life, but a large number allow them to experience vicarious violence. Games like Strike Force, Splinter Cell, Evil Dead, World of Warcraft, and Doom bring killing as an art form to children. There is hedonistic gratification in mass killing, and the child is rewarded by winning the game if he demonstrates sufficient levels of ruthlessness.

The joy of violence can also be found more indirectly in conquest and theft in popular culture. The recent film "The Wolf of Wall Street," directed by Martin Scorsese, shows a stockbroker who experiences intense joy by cheating other people and stealing from them, as well as from drugs and prostitutes. Making money from the naivete and ignorance of others by his hard sell approach caused a form of destructive ecstasy, clearly portrayed in the film, which worked against him as time went on.

In his book *The Anatomy of Human Destructiveness*, the psychoanalyst Erich Fromm outlines two categories of malevolent aggression. The first category concerns "spontaneous" forms of aggression, and the second one describes the malignant aggression which is "bound in the character structure." The "spontaneous" forms of aggression are most likely to occur when individuals or groups are subjected to a particularly extreme and traumatic set of experiences, and it can take the form of either "vengefulness" or "ecstasy." Vengefulness occurs when the individual or group has been subjected to intense and unjustified suffering, and becomes driven by the need to avenge their own pain. It is exemplified in the notion of blood revenge, "when revenge is a sacred duty that falls upon the member of a family, clan or tribe who has to kill a member of the corresponding unit if one of his people of his people has been killed."[51] Ecstatic destruction, though it is also a response to a particular set of circumstances, occurs in the ritualized orgies of violence or states of trance that certain societies use in order to give vent to extreme rage. Fromm writes that both of these forms of spontaneous destructiveness are responses to particular sets of circumstances. However, when the person's aggression is "bound in the character structure," it is particularly destructive, for it will then characterize the whole of the individual's way of life.

For Fromm, a central assumption is that each individual has his or her own particular "character." This term refers to "the relatively permanent system of all non-instinctual strivings through which man relates himself to the human and natural world."[52] Differences in character are rooted in differences in social experiences, both personal and cultural, although genetically given dispositions may also play a part. Some individuals, according to Fromm, experience so destructive an early environment, both personal and social, that the predispositions to behave in particularly destructive and cruel ways become bound in the character structure. Fromm gives as an example the sadistic character. The experience of absolute control over another human being creates the illusion of transcending the normal limitations of human existence, particularly for one whose real life is deprived of productivity and joy. He notes:

> For the sadistic character everything living is to be controllable; living beings become things. Or still more accurately, living things are transformed into living, quivering, pulsating objects of control. Their responses are forced by the one who controls them. The sadist wants to become the master of life, and hence the quality of life should be maintained in his victim. This is, in fact, what distinguishes him from the destroying person. The destroyer wants to do away with a person, to eliminate him, to destroy life itself; the sadist wants the sensation of controlling and choking life.[53]

Victoria Nelson discusses this form of projection as the "psychotopographic imagination," which externalizes traumas and psychoses onto the outer landscape. The externalizing of psychoses destroys any barriers that the ego might have for protection, and inner and outer worlds both mirror the projected threat.[54] It creates a sort of madness that is understood as ultimate truth, for there is no distinction between inner and outer worlds, and no way to disprove the threat. Sacred terror fuses with pathology, and the world is full of supernatural danger. There is no form of positive religious experience available that can balance the intensity of universal destruction. The person sometimes resorts to hatred and ecstatic violence to counter this projected threat.

Destructive ecstasy may exist in either a secular or religious framework, generally one where more positive forms of ecstasy are not available to the person. There are intense emotions like rage and fear and passion. The ecstatic desire to kill or harm may be found in the Roman arenas, with the crowds on their feet crying for blood, in fights where crowds scream at the winner to kill the loser, in riots and wars throughout the world. Wars do not always justify such destructive ecstasy—many people are horrified by the acts that occur, and are haunted by them for the rest of their lives. But others find their calling in wars, enjoying it as the ancient Babylonian goddess Anath was said to dance in rivers of blood.

Such intense negative emotion may also be found violating social norms, in choosing the forbidden—compulsive lust, theft, gambling, speeding, violence. It may sometimes be found in deliberately risking death in extreme sports, where the chance of death leads to exhilaration. When there is a greater the possibility of destruction the event becomes more attractive, thus the charm of demolition derbies and snuff movies, in which actors actually die in the violence during filming. There are currently "death races" for which one may register on the Internet, whose attraction is insanity and the possibility of death.[55]

The presence of intense, destructive emotion is perhaps the most obvious aspect of this "wargasmic" sort of ecstasy. Such intense emotion may also be found in conscious transgression, in violating social norms, and in choosing forbidden actions. In deliberately risking death, where rebellion and cheating death become goals in themselves, destruction becomes fascinating and a form of self-expression. It is a powerful mixture, the combining of *eros* and *thanatos*, the intense exhilaration of seeking, accepting, and combating the death urge.

Altered perception is less common in destructive ecstasy. It may come as a result of drink or drugs, when other people appear as enemies, when peaceful events are perceived as threats, and obvious lies are accepted and repeated as undeniable truths. It may occur during war, when violent actions no longer occur in a particular time and place, but are perceived as "out of time" or past wars which are echoed in the present. Modern wars can become sanctified by being understood as symbolic reenactments of ancient religious wars, and violence becomes acceptable because modern limits of "just war" are reinterpreted or suspended. There may also be individual cases of changed perception, such as battle frenzy, when the sense of time and space becomes altered.

In the religious world, we often have altered identity in a group context: We have Muslims fighting Westerners in Iraq who identify with Muhammad fighting the idol-worshipping Jahiliyah in Saudi Arabia, Christian fundamentalists claiming to be oppressed in the USA who perceive themselves as a part of the oppressed early church in Roman times, and Hindu fundamentalists still defending the armies of Rama in India against threatening Ravana. Alteration of identity is frequently found in destructive forms of ecstasy in vicarious enjoyment and identification with the aggressor. We can see this in boxing and other sports matches, where members of the audience identify strongly with the fighter and scream and yell and act out the actions that they want the fighter to perform. Such identification is encouraged in movies which are filmed through the eyes of the serial murderer, rapist, or torturer, allowing audience members to experience murdering or harming another person indirectly. It is seen in the propaganda of modern terrorist groups which encourages violent acts, which allows people to identify with the Jihadi online executioners who behead and disembowel others. Such merging of identity can reach toward a sort of possession, where the violence swamps the individual, making him or her part of a larger collective identity. Thus, we have suicide bombers identifying with the rage of their cultural group, soldiers being the wrathful hand of God, the kamikazi becoming the force of the destructive wind God or Kami, and the person who kills innocent victims because the Devil coerced or acted through him. The person is no longer an individual, but rather "a killing machine," possessed by the power of destruction.

However, we should note that transgressive violence often features an ambiguity about the roles of hero and villain, for the person who breaks the rules, fights against greater odds, and is exhilarated in the presence of grenades and firebombs can be either one, depending on the social circumstances. The force of salvation for one group can be the criminal and terrorist of another, when violence is justified by claims of oppression or group superiority.

Destructive ecstasy can be a problem for both individual relationships and for society as a whole. Yet people are drawn to ecstatic states, and if a society is threatened by and rejects the more positive forms of ecstasy, it may then suppress or outlaw both types. Then, the more loving and creative forms of ecstasy become confused with the destructive ones, and because they are all illegal they then become interchangeable. Visions of light and visions of darkness are both forbidden and thus are equated. It is no longer an opposition between good and evil, but between ordinary and non-ordinary. Ecstasy itself becomes a threat, and then society lashes out at anything understood as abnormal. Religious varieties of ecstasy, even those that emphasize love and unity, are made illegal, immoral, and profane, and all forms of ecstasy become equated with transgression.

The darker side of mystical ecstasy is usually portrayed by scholars as an uninvolvement in earthly ethical problems, in favor of a monistic unity which has no room for a dualism of good and evil. In the book *Crossing Boundaries: Essays on the Ethical Status of Mysticism*, it is debated whether mystical and ecstatic states can separate people from the world and its values, making people variously moral, amoral, immoral, or transmoral. The "antinomian challenge" is the tendency for unusual states to lead to moral problems, such as life-negation, "effeminate weakness" on the part of the mystics, extremism, and abuse. A concern with spiritual knowledge rather than social ethics encourages future bad behavior; as writer Jeffrey Kripal notes, for in his view mystical experience "inevitably tends toward the transgressive, the ascetic and the extreme."[56] It might thus be in the interest of cultures and religions to forbid ecstatic experience, to be protective, to help people who "suffer" from visions and ecstasies.

However, by doing so, the culture may not be helping either itself or the ecstatic "victims." By forbidding subjective experiences understood as deeply meaningful, societies and religions may well create the problems that these writers seek to prevent. When ecstatic states are discouraged, and societies emphasize rigid authority and obedience to particular belief systems, they virtually guarantee that at least some people will not fit into their social structures. It is useful to examine the consequences of such a critique, which results in limiting access to ecstatic states. As the ethical narrows the mystical, some ecstatic experiences will react against that pressure. Kripal notes that transgression is the key or linchpin to the debate about the ethical status of mysticism.[57] As we expand the realm of the forbidden, the potential for transgression grows ever larger.

From this perspective, without strong moral control and punishment, what we would have is immorality. This is certainly a traditional, conservative fear. To bring this to its extreme, when moral rules cease to dominate thought and behavior, people will choose evil rather than good. Ecstatic religious experiences "inevitably" lead to weakness or suicide or malice. It is a particularly dark view, which fits in well with a notion of fallen humanity. However, it is not only ecstasy which leads to problems—as we have seen, lack of ecstasy can evoke the same problems. As the writer Novalis stated in his novel *Heinrich von Ofterdingen*, "Where there are no gods, ghosts will hold sway." If there is no positive goal, destructive ones can take its place.

Perhaps we need a better model for what is acceptable. As G. William Barnard notes, it is not the obligation of society to force all people into a single mold, and condemn all behavior outside of that narrow range:

> Perhaps the universe does not just need social reformers, but a whole spectrum of people responding in different ways to their own unique gifts and inclinations. Perhaps it is the job of enlightened beings to energize, enliven and inspire those who come to be with them, and this task is worthy of tremendous respect on its own merits.[58]

Allowing a wider spectrum of states and roles as acceptable might also allow for greater, positive exploration of ecstatic gifts. If societies had less fear of transgression, and emphasis upon it, we might find fewer transgressive acts as responses to the limitations of religion and culture.

One response to the pathologizing and condemnation of more positive ecstatic states is the rise of destructive forms of ecstasy. Another is the creation of alternative forms of ecstasy, when ordinarily secular acts come to take on religious language, interpretations, and goals. We can see this in the next chapter.

Notes

1. Kingsley Browne, *Co-ed Combat: The New Evidence That Women Shouldn't Fight the Nation's Wars* (New York: Penguin, 2007), p. 115.
2. Victoria Nelson, *The Secret Life of Puppets* (Cambridge: Harvard University Press, 2001), p. 19.

3. Ibid., p. 9.
4. Ibid., p. viii.
5. Lars Bang Larsen, "When the Light Falls: Notes on Ecstasy and Corruption," in *Ecstasy: In and About Altered States*, ed. Paul Schimmel (Cambridge: MIT Press, 2005), p. 176.
6. Ibid., p. 178.
7. Mihaly Csikszentmihalyi, *Flow: The Psychology of Optimal Experience* (New York: Harper and Row, 1990), p. 69.
8. William F. May, "Terrorism as Strategy and Ecstasy," *Journal of Social Research* 42 (Summer 1974): 285–287.
9. Robin Morgan, *The Demon Lover: The Roots of Terrorism* (New York: Simon and Schuster, 2001 (1989)), p. xvi.
10. From an October 7, 2001, *New York Times* article, cited in Morgan, ibid., p. xxi.
11. August 1993, cited in Morgan, ibid., p. xxxiii.
12. Ibid., p. 23.
13. Ibid., p. 63.
14. Ibid., pp. 229–230.
15. William Broyles, "Why Men Love War," online at William Broyles, Jr. "Why Men Love War," Esquire, 1984, online at http://www.wsu.edu/~hughesc/why_men_love_war.htm.
16. Ibid.
17. Ibid.
18. Discussion online at http://www.amazon.com/WAR-Sebastian-Junger/dp/0446556246.
19. Michael Eigen, *Ecstasy* (Middletown: Wesleyan University Press, 2001), p. viii.
20. Ibid., p. 2.
21. Ibid.
22. Ibid., p. 6.
23. Ibid., p. 9.
24. The term "ultraviolence" comes from Anthony Burgess' book *A Clockwork Orange*, which later made into a controversial film. The term refers to a character's positively conditioned responses to feelings of evil and his love of random, orgiastic violence.
25. This description of the book comes from the advertising by SUNY Press. Murray Pomerance, ed., *Bad: Infamy, Darkness, Evil and Slime on Screen* (Albany: State University of New York Press, 2009).
26. Citation from "Interview with a Vampire" in Ray Broadus Browne and Gary Hoppenstand, eds., *The Gothic World of Anne Rice* (Madison: University of Wisconsin Press, 2009), p. 231.
27. David and Ellen Ramsdale, *Sexual Energy Ecstasy* (New York: Bantam Books), p. 257.

28. Jenny McCartney, "The Rise of 'Torture Porn'," *The Week*, July 27, 2007, p. 36.
29. Peter Bebergal, "A Meditation on Transgression: Foucault, Bataille and the Retrieval of the Limit," online citation www.ctheory.net/text-file.asp?pick+102, p. 1.
30. Bataille, *Oevres Completes*, IV, 23. Cited in Adrian Gargett, "Vampire: Georges Bataille and the Philosophy of Vampirism," in *The Richmond Review*, online at https://richmondreview.co.uk/features/gargett01.html, p. 2.
31. Ibid., p. 3.
32. Bataille, *Oevres Completes*, V, 264, in ibid., p. 8.
33. Ibid., III, pp. 11–12.
34. Ibid., IV, p. 219.
35. "On Suicide Bombing" by Talal Asad, at http://www.seagullindia.com/booksblog/post/From-On-Suicide-Bombing-by-Talal-Asad.aspx.
36. Foucault's ideas described in Peter Bebergal's "A Meditation on Transgression: Foucault, Bataille and the Retrieval of the Limit," ibid., p. 3.
37. Dan Stone, "Genocide as Transgression," *European Journal of Social Theory* 7, no. 1: 45.
38. Saul Friedlander, *Memory, History and the Extermination of the Jews of Europe* (Bloomington: Indiana University Press, 1993), p. 99.
39. http://www.culturenorthernireland.org/article/1304/cuchulainn-and-the-t-in-epic.
40. Stephen R. Lawhead, *Merlin: Book Two of the Pendragon Cycle* (New York: Avon Books, 1988), p. 231.
41. Anonymous, *The Ecstasy of Pain: A Book on Flagellation, Remarkable Cases of Flogging and of Celebrated Flagellants* (New York: Brown Book Company, 1962), pp. 27–28.
42. Ibid., p. 34.
43. Daniel Benjamin and Steven Simon, *The Age of Sacred Terror* (New York: Random House), pp. 91–93.
44. See http://www.inter-islam.org/faith/Signs-Of-Qiyaamah.htm.
45. Ibid., p. 92.
46. Ibid., p. xx.
47. Yaakov Lappin, International Terrorism: Terrorism 2.0, online at http://www.jpost.com/VideoArticles/Article.aspx?id=192353.
48. The film "Iranium" is described in the article "Iran can't intimidate Canadians" by Clifford D. May, *The Post and Courier*, January 28, 2011, p. 13A.
49. Cited in Thomas L. Friedman, *The Lexus and the Olive Tree* (New York: Farrar, Straus and Giroux, 1999), pp. 327–328.

50. Ibid., p. 200.
51. Erich Fromm, *The Anatomy of Human Destructiveness* (New York: Penguin, 1977), p. 363.
52. Ibid., original italics.
53. Ibid., p. 388.
54. Ibid., p. 110.
55. As an example, see www.youmaydie.com.
56. *Crossing Boundaries: Essays on the Ethical Status of Mysticism*, ed. G. William Barnard and Jeffrey J. Kripal (New York: Seven Bridges Press, 2002), p. 33.
57. Ibid., p. 69, note 118.
58. Ibid., p. 84.

CHAPTER 6

The "Spiritualized" Ecstasies: Sex, Drugs, and Rock and Roll

Another set of ecstasies might be called "spiritual" ecstasies, the modern transformation of the old trinity of sin: sex, drugs, and rock and roll. These used to be widely condemned by Christian preachers as sinful acts which caused the Devil to live in the flesh, and would open the gates of hell to those who indulged in them. It was the job of evangelists to rail against such sins, to keep good people from getting involved in them.

Now, the government has stepped into the place of the preachers—we have billions of dollars spent on Abstinence-Only sex education, the national and international War on Drugs, and regulation of the music and entertainment industries through censorship. The government spends more money than church pastors and tent revival preachers ever did, yet sin continues to thrive.

An alternative approach to this set of sins has been taken by some New Age groups. Instead of being paths to hell, they are really stairways to heaven. By doing such actions in the right manner, they are rescued from being sinful indulgences and turned into virtuous acts. Thus, we see sexuality, formerly understood as promiscuity and hedonism, now reinterpreted as tantric practice, the body's way to God. Sexual ecstasy becomes religious ecstasy, for God dwells in the body. Drugs, popularized as causing anything from addiction to demon possession, are now sacraments, with peyote, ayahuasca, and marijuana used in rituals to contact an immanent divinity. Rock and roll, the uncontrolled music and

J. McDaniel, *Lost Ecstasy*, Interdisciplinary Approaches to the Study of Mysticism, https://doi.org/10.1007/978-3-319-92771-8_6

dance which was accused in the early twentieth century of being the "Devil's music," has become a part of raves and music festivals, bringing group harmony and ecstatic joy to thousands of people at a time. When the mainstream religions close their doors to religious ecstasy, people seek and find it in other places.

Do these transformations have any historical precedents? In India and Tibet, we see occasional uses of ritual sexuality as an aspect of the tantric traditions, with its primary emphasis as mystical union with deities, gaining long life or immortality, and the development of supernatural abilities (or *siddhis*). Instead, New Age writing tends to emphasize the goal of ecstatic sexuality as the healing of sexual traumas and making the person healthier and stronger. There are many indigenous traditions around the world that make use of drugs, requiring religious experts who undergo extended initiations, and use the drugs to gain spiritual insight for the sake of the community. We do not see such drugs used for personal entertainment, as we see today. As for rock and roll, ecstatic music and dance are seen in many traditions in Africa and the Caribbean, including the possession dances of Santeria and Vodou, and in healing and shamanic dances in India and Southeast Asia. Today's ecstatic music rarely causes possession trance, though it may induce group enthusiasm and *communitas* at raves, or focus upon deities of different religious traditions, as in the Dances of Universal Peace.

The umbrella terms "New Age spirituality" and "Spiritual but not Religious" cover a variety of different belief systems and practices. In these cases, the term "spiritual" focuses on the experiential, usually without a clearly defined Theology. Rebelling against the authoritarian hierarchies and dogmatic demands of their pasts, people seek a relationship with a loving divine entity that doesn't demand things of them and look down in disapproval, or a relationship with the world that is naturally harmonious. According to a study conducted by Pew Research Center in 2012, the number of Americans who do not identify with any religion has increased from 15% in 2007 to 20% in 2012, and this number continues to grow. One-fifth of the US public and a third of adults under the age of 30 are reportedly unaffiliated with any religion, yet they identify as being spiritual in some way. Of these religiously unaffiliated Americans, 37% classify themselves as spiritual but not religious, while 68% say they believe in God, and 58% feel a deep connection to the earth.[1]

Spiritual ecstasies mix old and new shamanic trances and yogic mantras with new models of the mind and neurobiology in a consciously chosen syncretism. As modern Western religion becomes ever more rational and ethical, incorporating new scientific and psychological models as means of explication and justification, New Age spirituality goes in the opposite direction and becomes more archaic and mythical. The ecstatic states associated with sex, drugs, and popular music that have been rejected and described as "of the Devil" by the mainstream churches and synagogues have found a new home.

Connecting Sexual Ecstasy with Spirituality: From Tantra to Romance Novels

The first alternative spirituality to be examined is sexuality as a path to divinity. Next to the drug Ecstasy, sexuality was the most popular topic associated with ecstasy by colleagues and the media. It seems that, as religious ecstasy has become inaccessible or disappeared from the modern discourse, sexual ecstasy has come to take its place. In some cases, sexual ecstasy is spiritualized, and in others, it becomes a stepping stone to a deeper and more meaningful spiritual ecstasy.

Sex, with its resultant ecstatic state, is used to advertise everything from cars to beer, and its imagery is ubiquitous in the worlds of sales, entertainment, and even sports. From a topic which could not be mentioned in proper Victorian society, it has gone on to be discussed endlessly in talk shows, therapy groups, men's and women's magazines, and in relation to such legal drugs as Viagra, Levitra, and Cialis. In advertising, sex becomes a competitive sport, with people competing for the biggest, best, and longest orgasms (and with drug influence, for those at the most advanced age). We also see the rise of vicarious ecstasy, through sexually explicit films and Internet pornography.

Sexual ecstasy is probably the most popular and widespread form of ecstasy in the world. It does not require religious or mystical knowledge, or any particular belief system. But can sexual orgasm really be called religiously ecstatic? It is described that way by metaphors in popular culture. In this book, our definition of ecstasy is intense emotion, altered perception, and an altered sense of identity in relation to an ultimate goal (for religious or mystical ecstasy, in relation to a specifically religious object). According to this definition, we could say that orgasm is ecstatic insofar as it involves intense emotion and altered perception. It would be difficult to achieve orgasm with total disinterest. It fits the definition of ecstasy most clearly under altered perception, as the nerve stimulation brings perceptions that people do not otherwise have. Orgasm may occasionally evoke an altered sense of identity, in those events in which people feel themselves merging into each other or "becoming one flesh," and in those cases where people temporarily lose any sense of personal identity. Thus, we could equate ecstasy and orgasm to a limited degree, primarily in the areas of perception and emotion. It becomes closer to religious ecstasy when the sense of union has a theistic or transpersonal focus or goal, i.e., a religious object.

We begin with a look at Western Tantra. Tantra is a meditative and ascetic tradition that has been developed and elaborated in India, China, Tibet, and other countries in Asia. It involves ritual texts, contemplative practice, cosmologies, a knowledge of sacred languages and mantras, the use of hand and body positions, visualizations, and meditation on deities. Occasionally, it involves rituals of sex and death, the two greatest sources of emotional attachment that must be overcome by the spiritual seeker, and it may focus upon a single goddess or upon a divine couple. Participants seek liberation from illusion, or a close relationship with a deity, and in some cases supernatural powers. This may be attained through yogic detachment, or through identification with deities. The forms of tantra that we see in the modern West bear little relationship to the older traditional forms of tantric belief and practice.

Tantra in the West has largely lost its theistic, ascetic, and contemplative dimensions. It may emphasize the energy of the body or cosmos, or a generalized goddess figure—the tantric gods have disappeared or become strictly symbolic. While tantra in India has both devotional and folk dimensions, tantra in the USA and Western Europe largely ignores these and tends to have one dimension: that of ecstatic hedonism. Rather than giving up attachment to the world for the sake of the goddess, here we have people immersing themselves in the world to find her. She has become immanent rather than transcendent, associated with self-satisfying earthly lust *(kama)* rather than self-sacrificing spiritual love (*prema*).

Religions change as they travel and become popularized, and thus, there are many schools of Western Tantra. Probably, the major popularizer of tantra in the West, and the origin of many of these tantric offshoots, was the late guru Bhagavan Rajneesh. He was born in 1931 of Jain parents, into a religion which emphasizes asceticism and nonviolence. After an early experience of being "mad with blissfulness," he rebelled against Jainism, calling its ascetic tradition masochistic, and he emphasized the value of sexuality that Jainism believed in regulating and suppressing. He came to condemn all organized religions as dead, life-denying, and perverted in their rejection of the physical world. By the late 1960s, he began to develop his system of Dynamic Meditation, a state of induced chaos which he said would free the person to experience intense joy and cleanse the body of its anxieties and stresses. He called his followers Tantric neo-sannyasins, who were not rigid and life-denying like traditional renunciant sannyasins, but rather "joyous creatures, rebels and dancers." He was the Tantric Master, making sexuality the spiritual path of his disciples.[2] As he believed that Westerners cared most about sex and money, these were the emphases for his Western Neo-tantra. Rajneesh changed his name to Osho and continued to experiment with meditation and encounter groups until his death in 1990. As he defined tantra, it meant gaining cosmic consciousness.

> Tantra means expansion. This is the state when you have expanded to the uttermost. Your boundaries and the boundaries of existence are no longer separate, they are the same. Less that that will not satisfy. When you become universal, you come home.[3]

Rajneesh's Western Tantra encourages sexuality:

> The greatest, most intelligent people are the most sexual people. This has to be understood, because love energy is basically intelligence. If you cannot love you are somehow closed, cold; you cannot flow. While in love one flows. While in love one feels so confident that one can touch the stars. That's why a woman becomes a great inspiration, a man becomes a great inspiration. When a woman is loved she becomes more beautiful immediately, instantly! Just a moment ago she was an ordinary woman... and love has showered upon her, she is bathed in a totally new energy, a new aura arises around her. She walks more gracefully, a dance has come to her step. Her eyes have tremendous beauty now, her face glows, she is luminous. And the same happens to the man.[4]

However, his followers must not cling to one partner, for that simply causes more bondage. God has appeared in millions of forms—why cling to one? What is desired is both freedom and experience, which comes through expanded sexuality. As one of his poems states:

> Cosmic orgasm
> is just like
> two lovers making love
> but a millionfold,
> because now you are
> making love with
> the whole universe.
> That is why Tantra is
> known as the yoga of sex,
> the path of love.[5]

Rajneesh's tantra was a rebellion against organized religion, emphasizing free sexuality and ecstatic spirituality. It has generated and influenced a variety of other Western tantric groups.

Another modern form of tantra is Oceanic Tantra. Oceanic Tantra emphasizes "dolphin-breath and undulatory movement" as a means of generating and manipulating sexual energy, to gain health and longevity, and show "ways [that] men and women can be multi-orgasmic." Various workshops and seminars make use of dance, music, theater, visualization, "love in the sensuous waters," and performing breathing and wave movements which generate "interspecies communication with dolphins and whales" (in order to feel the "tantric pulsation of the universe"). Begun by a musician and a psychotherapist at their Kahua Hawaiian Institute, the goal of this form of tantra is to create "passionate and dynamic ways of bringing greater consciousness into the areas of sexuality and spirituality." The goal is divine unity with a religious goal, in this case one's own divine aspect.

> As sexual energy rises it can rejuvenate every cell in our body. By connecting with the heart it becomes prayer; rising up into the mind we can experience visions and revelations; and ultimately we can experience communion with the divine. This fulfills a deep yearning in all of us to experience our own divinity.[6]

Another tantric organization, the SkyDancing Institute, advertises its "Tantra Love and Ecstasy Trainings." As a brochure's introduction states, "Our activities provide a gateway into the Tantric way of life by helping you to awaken to your ecstatic potential, embrace your own vital powers, create a more fulfilling love life and to enliven your sensuality as well as your spirituality." SkyDancing tantra claims to integrate "the ancient spirit of Tantra" with humanistic psychology and clinical sexology, to present the Tantric Paradigm. There are workshops in which sexual energy is used as a magical tool; people visualize their desired situations, and the visualized image is charged with "orgasmic energy" in order to manifest it in the world. Their brochure claims "scientifically documented benefits."[7] SkyDancing tantra is primarily taught by workshops and short-term training seminars, either by its originator, Margo Anand, or by teachers who have been certified by the program. It leads to both ecological and mystical insight:

> The ecology of the body is the same as the ecology of the planet, and both are the same as the ecology of the universe. In the experience of ecstasy, we are linked to the cosmos and the cosmos is within us. In this way we understand that we are divine and that there is no difference between God/Goddess and each of us.[8]

According to various Western tantric schools, people can gain salvation from all sorts of things, from boring lives to traumatic wounds, through orgasm. Such a salvific orgasm cannot be an ordinary one—it must be endless, eternal, part of the deepest aspect of the natural world. Such orgasms are called valley orgasms, and while they may come about by accident, there is a burgeoning collection of meditative techniques to generate them. There are practices inducing extended orgasm and multiple orgasms, involving breath control, exercise, visualization, and occasionally biofeedback. Most practices combine meditation and erotic activity.

A good example of these practices is given by writers David and Ellen Ramsdale. They suggest redirecting the energy of the orgasm to the top of the head, so that the person feels like the orgasm is occurring through the crown of the head. A fountain of clear bright white light should appear to rise up, and the person should feel him or herself floating up through this fountain and flying in the air. The light should then shower down "like a blissful rain of unspeakable beauty and kindness," a waterfall of contentment pouring down in waves.[9] There is also the Breath of Fire, a practice of ecstatic orgasm in which the person visualizes the chakras being charged with energy, from the lowest to the highest chakras. If one visualizes these along with tightening various muscles, ecstatic energy will arise which causes moaning, trembling, and shaking wildly.[10] Visualizations which increase the ecstatic power of the orgasm include having the person imagine the body as a garden, light circulating through the body, orgasm as a chain of exploding firecrackers, chakras emitting red light like a ruby laser, and the partners' bodies melting like butter on a hot stove and dissolving into each other.[11] The Ramsdales also describe Metasexual Lovemaking, in which the body is sensitized so that orgasm can occur in such unlikely locales as the thumb, the elbow, the eye, the toe, or the heart. These can be associated with music—they suggest Ravel's Bolero, or Tchaikovsky's 1812 Overture.[12] They also note that such "consciousness-raising sexual activity" (as they define tantra) has many variants, including Polynesian, Egyptian, Scandinavian, and African.[13]

The Ramsdales note that another way to induce sexual ecstasy is consciously combining sexual activity with religious imagery. One may imagine having sex with Jesus Christ, Mary Magdalene, the Buddha, Kwan Yin, Radha, Krishna, Kali, or some other deity or saint.[14] Or one may watch a picture of the deity or saint during sex, and imagine a relationship with him or her. From a different perspective, the couple may imagine that they are different kinds of animals mating, or that their bodies are changing into different bodies, or bodies of different colors. Depending on the worldlier inclinations of the people involved, making love on piles of money may also cause ecstasy. One practice that the authors seem to prefer is one that they call the Wave of Bliss. In this practice, life becomes a sort of endless ecstatic sexuality in relation to an infinite universe, an eternal and repeating cycle of foreplay, arousal, orgasm, and afterglow.[15] The practitioner should be totally involved in the cycles of expansion and contraction, rising and falling, for life becomes intense and ecstatic when it follows these rhythms and cycles. The goal is to intensify or divinize orgasm, combining sexual ecstasy and religious experience.

Tantric practice may also be combined with positive thinking, as orgasm is considered to be a good time for affirmations. These affirmations may have spiritual goals, or more worldly ones. The authors suggest using ecstatic energy toward gaining health, wealth, love, and losing weight. This approach of affirmations or visualizations at the moment of orgasm is also seen in the Western magical tradition, in which ecstasy and magical power occur together. Wiccan practitioners may dedicate their orgasms to healing other people, to changing the world, or to bringing about a desired situation.

These "spiritualized" forms of sexuality generally do not emphasize reproduction. The goal is to enter an ecstatic state which will give religious insight into deities and send blessings to the group. However, there are spiritual paths which unite religious ecstasy and conception.

The Unification Church, begun by Sun Myung Moon, has interpreted sexuality as an action that can be used for degeneration or purification. Mankind was corrupted by a fall long ago, but this was not due to Eve eating an apple. It was due to her having sex with the Snake, who was really the Devil in disguise. All humanity is descended from that mating, and all people therefore have "snake genes." These give people impulses toward evil and are inherited. The only way to have perfect and pure children is to visualize Sun Myung Moon and sense his presence at the moment of conception. This ecstatic vision will transform the corrupted snake gene into a perfect human gene. Only by having a race of perfect people, conceived while visualizing the Lord of the Second Advent (the true role of Sun Myung Moon) will humanity reach its full potential and become a race of ideal people. As the Lord of the Second Advent, Moon is God returned to human form to generate a race of perfect human beings. The Lord of the First Advent (Jesus Christ) did not succeed in his mission to perfect humanity and instead was crucified. So Moon will succeed at the mission at which Jesus failed, and by perceiving him at the moment of conception, divinity fuses with sexual ecstasy, and genes will be transformed to create a new race of ideal people.[16]

In this way, the presence of a deity blesses the sexual act and any children resulting from it. It is similar in some ways to the kabbalistic practices intended to give birth to the Messiah, which also involve ritual purifications and visualization at the time of conception. While the idea was widespread in medieval Europe that a woman would bear more perfect children if she saw the image of Virgin Mary or a saint during the process of childbirth, it is less usual for the imagery to be used at the moment of conception.

These rituals are an attempt to reverse traditional teachings about sexuality in the West. Since the patristic period of Christianity, sexuality was considered to be sinful, only allowed within marriage, and then only for reproductive purposes. From the fifth century onwards, with the writings of St Augustine of Hippo, original sin was believed to corrupt every aspect of life. Sin was an "infection of nature," which showed itself through the presence of concupiscence or insatiable desire. While the early church emphasized celibacy as the most desirable state, and virgin martyrs and saints were glorified, eventually marriage became a sacrament and was theologically justified by procreation. Because sin had corrupted human passion, people should never have sex for pleasure alone.[17]

Sexuality came to be understood through a Theology based on natural law, God's law showing itself through nature. Thus, for instance, masturbation was considered to be a greater sin than adultery or fornication by St. Thomas Aquinas, because it violated the laws of nature as well as society and the church. Passion corrupted reason and required contrition; Pope Gregory the Great said that married people should not receive communion after sex unless they had first done penance.[18] There has been a long history of Christian writings on the dangers of sexuality, especially when it escaped institutional control. Though the Song of Songs had been accepted into the Biblical canon, and sensuality was generally seen in a positive light in Judaism, it has been widely condemned in Christian history. Most Catholic saints and ecstatics were celibate—the passionate imagery of love and marriage with Christ was quite opposed to their lives of fasting, prayer, and contemplation. Protestant Theology also followed Paul, in warning against the sin of lust and the dangers of unmarried intimacy. Ecstasy and physical sexuality could not be easily harmonized in a tradition that emphasized sin and the evils of the flesh.

The nineteenth and twentieth centuries brought challenges to these views, but they came slowly. One bastion of the older suspicion of fleshly evil was the field of medicine. We see medical writings from the nineteenth century that describe how orgasm in the wrong social situation caused epilepsy, insanity, and crime (some ministers added leprosy and tuberculosis), and that masturbation caused weakness (as the loss of one ounce of semen was equivalent to the loss of forty ounces of blood). As one medical writer said, the "venereal paroxysm" swept along the nerves "with the tremendous violence of a tornado," with the "convulsed heart" driving blood into other organs and damaging them. Of course people would be weakened by it:

> The nervous system, even to its most minute filamentary extremities, is tortured into a shocking state of debility… the muscles, generally, become relaxed and flaccid; and consequently, all of the organs and vessels of the body, even to the smallest capillaries, become exceedingly debilitated; and their functional powers exceedingly feeble.[19]

This is what comes from sexual ecstasy, even if only by "adultery of the mind."

However, there were also more liberal Christian writers, who emphasized the goodness of creation and sexuality as part of God's plan for mankind. Thus, sexuality becomes marked by the absence of guilt and is even understood as healthy—it is no longer only a sign of mankind's fallen nature. Indeed, sex could be a spiritual act, blessed by Jesus, who set it free from sin.[20] Of course, such joy was only allowed within a situation of love and marriage. The idea of sexuality as valuable, even as a form of salvation (especially within a monogamous romantic relationship), has slowly been showing up in sex manuals and Christian marriage therapy through the twentieth century. These justify sexual activity as acceptable behavior against the older background of theological and Victorian disapproval.

But the salvific aspect of sexuality has probably been more widely distributed through romance novels which present an idealized mixture of romance and implied sexuality which represents an interpersonal rather than a spiritual salvation. Some Christian novelists, seeing the popularity of these novels, sought to take this romance novel genre and add God and spiritual love, associating romantic ecstasy and its sexual component with Christian spirituality.

Both the older mainstream romance novels and the recent Christian ones bring ecstasy back into human relationship—but primarily through romantic love rather than sexual activity. Ecstasy is the glue that binds man and woman through love, and through love they realize God's presence. But the focus is ecstatic love rather than orgasm, and in the Christian romances, *agape* rather than *eros*. It is love that shows the presence of God, experienced vicariously by the reader throughout the text.

In many secular romance novels, ecstasy is a sort of cathartic bliss, for the beautiful but lonely virgin desperate to find love, and the cynical rogue trying to find lost idealism and innocence once again. Modern romance novels tend to follow set formulas, with a variety of locales and characters. According to one formula, we tend to see large, strong men—warriors, aristocrats, explorers, ship captains—in positions of authority and power. We also see sensitive, lovely, yet perky and spunky young women, who are courageous and often rebellious. The heroine is usually young, and often in trouble, but she perseveres and eventually there is a romantic, positive ending. The relationship between the hero and heroine of the novel is rarely easy—there are tensions, uncertainties, anxieties. Yet, all is solved in an ecstatic embrace, when a passionate kiss brings hidden or forbidden love to the surface.

The romantic hero realizes that his beloved is unique; she realizes that he is in love with her, and a good man after all. The ecstasy is the realization of the true nature of their relationship. In more modern novels, we have a few additions—such as plots where involuntary sex brings realization of desired sex, and where the sexual act (which was due to plot twists rather than malice or simple lust) bonds them and makes them soul mates for life. The relationship between two "souls" implies spiritual depth, but there is no clear religious object in the secular ecstatic mix.

For the reader, it is a sort of vicarious ecstasy, in which the simulation of imagined passions of fictional characters is intended to bring forth imagined states of passionate love in the readers. As a sample of sexual ecstasy in romance novels, here are a few sentences from Elizabeth Lowell's book *Tell Me No Lies:*

> He saw her watching him, smiling in the knowledge of his release even as ecstasy swept through her again... He had not breath, no thought, no sight, nothing but the ecstasy consuming him, and he wondered if this was what it felt like to die...
>
> The sound of her name breaking on Catlin's lips hurled Lindsay into ecstasy... Lindsay clung to Catlin, absorbing the shock waves of his ecstasy into herself...[21]

Sexual ecstasy is secondary as a means of expressing what is primary: ecstatic love. As Barbara de Angelis notes in her book *Passion*:

> The heart is the home of true passion. There, in the mysterious, invisible realm from which all feeling emerges, the ocean of passion lies waiting to be stirred, not by the right touch or a skillful caress, but by the one thing that can cause passion to rise up- the love from another heart... it is the love which emerges when you are able to give yourself completely to another, a passion that turns mere connection into union, and mere physical pleasure into emotional and spiritual ecstasy.[22]

For secular romance novels, love is the ultimate concern, and realization of love brings ecstatic joy. However, in religious romance novels, the hero is often idealistic—he may be a doctor, a missionary, a man in seminary wondering about his vocation. The couple's mutual attraction is based on their shared faith in Jesus. Often the woman brings the man to renewed faith, and he gives her a ring in return. Women may act as saviors, bringing faith to men who had doubted, but have now gained new inspiration for life. The men may then bring their faith to the world. Ecstasy comes through the kiss of commitment, the woman's realization of her absolute dependence on the man, and for both the realization of their absolute dependence upon God. Sexuality might come later, but it is not the source of greatest ecstasy. Religion and romance in these novels are linked through ecstasy, and that link has a long history, as Diane Ackerman notes, "Religious ecstasy and the ecstasy of lovers have much in common-the sudden awareness, the taking of vows, the plighting of troths, the all-consuming fire in the heart and flesh, the rituals leading to bliss…"[23]

While such love may represent a "spine-tingling, rip-roaring, drug like addiction," it may also have elements of vulnerability, fantasy, and the excitement of "galloping a horse in a lightning storm." Ecstatic religion and romantic love share the desire of fusion with the beloved, following Plato's image: "lovers are incomplete halves of a single puzzle, searching for each other to become whole… at some point, all lovers wish to lose themselves, to merge, to become one entity. By giving up their autonomy, they find their true selves."[24]

This romanticized approach comes partly from the older ideas of Catholic romantic love, in which celibacy is an offering to Jesus, and virginity and purity reflect the individual's virtue and piety. It also includes ideas of mystical union portrayed through romantic love of Jesus as bridegroom. In marriage, sex is sacramental, a foretaste of heaven, given by God to bless the responsibility of parenthood. It is participation in divine creation, which is shared with God, and creating a child in the physical world is like the creation of mountains and oceans at the beginning of time. A sanctified union is creative and brings parenthood. In the ideal Christian relationship, the woman is submissive, and the man is respected.[25] There is dependence by the woman on both the man and upon God.

It is this sort of dependence, glorified in the popular literature and in the mainstream religions, which became understood as a limitation by the liberation movements of the twentieth century, for both the Women's Liberation Movement and the Sexual Revolution. In the older approach, the woman's ecstasy was entirely dependent on the man, as was the woman herself. If he grew violent, or hostile, or otherwise unsuitable, the religious woman was doomed to a life of misery, with neither ecstasy nor personal freedom. The limitations of such a situation motivated women to take responsibility for their own sexuality—and eventually, for the meaning of their own lives. They sought not to be dependent on a husband and a religious institution. This was dramatized in Eve Ensler's *Vagina Monologues*, which examined many women's approaches to sexuality. In one monologue, a woman whose orgasms were erratic and unpredictable learned that she could influence, and even control, when they occurred. One does not undergo ecstasy, and one consciously achieves it. Ecstasy becomes less religious, and more incorporated into daily life, a product of action and will.

This more feminist view of sexuality as being free, an act whose goal is interpersonal bonding rather than procreation, has long been opposed by set notions of sexual morality and immorality, and the importance of those with the moral high ground regulating the sexual behavior of others. But understandings of ecstasy may differ. If the goal is sexual emancipation and freedom, ecstasy will be emotional and biological. If the goal is having sexuality bound to institutional norms and religious ideals, ecstasy will be understood as rationality (in conformity with God's will) and obedience, an expression of faith contributing to the building God's kingdom. If sexuality becomes a part of God's plan on earth, we can have approaches like Matthew Fox's Creation Spirituality, in which the focus is upon the creation of the world and all things in it as good. This approach argues against the Fall and Original Sin as the most significant Biblical acts for mankind. Sexuality thus becomes a form of worship, and its ecstatic goal brings one into close relationship with God.

We might put ecstatic sexuality into three categories. It can be a competitive sport, whose goal is spiritual transcendence. Here, we have the Western varieties of tantra, with valley orgasms and multiple orgasms as a path to spiritual success and to experiencing the deeper, hidden yogic energy centers in the body. If these do not bring the person to ultimate and transpersonal states, it is likely because of "sexual wounding," or previous insufficiently ecstatic sexuality. There are workshops to overcome such limitations—for the more orgasms you have, the closer to God you will be. It can induce altered states of consciousness, and even religious visions and cosmic union, as a result of a conscious process of transformational practice. The spiritual object in this form of ecstasy may be symbolized as "the goddess," the cosmos, or a "life energy" within or beyond the physical body.

A second category is sexual ecstasy as sacred offering, in which orgasm is a part of the ritual where the individual acts as a magician capable of changing the creative or causal flow of the universe. In the understandings of sacred marriage in some types of Western Tantra, Wicca, and the

Western magical tradition, there is a transfer of merit and power through orgasm. If the person thinks of a goal, especially a spiritual one, at the moment of orgasm, that goal can be achieved. Orgasm becomes an offering to the deity, and the shared sexuality can evoke the presence of the goddess or the divine couple.

A third type of ecstatic sexuality focuses on an altered identity in which the person (or a couple) becomes identified with a god, and thus, sexuality is by definition a relationship with the divine. One partner may be understood as an incarnation of a God or possessed by a deity, or the couple may visualize themselves as a divine couple (Shiva and Shakti, Jesus and Mary Magdalene, Radha and Krishna). There would be a great discrepancy of power if only one member of the couple is a deity, but dominance and submission can become a part of the ritual. A lesser variation of this is having the couple, or one member of the couple, blessed by a god, and these blessings can be shared through sexuality.

Thus, sexual ecstasy may be understood in both secular and religious or spiritual terms. It is capable of producing intense alterations in identity, perception, or emotion while being freed from its older associations with sin and damnation. We see the same secular-spiritual dichotomy in our next area of study, the relationship of psychedelic drugs and ecstatic states.

Drugs as Entheogens: Bringing the God Within

In the modern West, ecstasy is not usually associated with religion in the popular imagination. It is associated primarily with drugs (especially the drug MDMA, commonly called "Ecstasy"). The condemnation of ecstasy in the modern West comes partly for doctrinal reasons—the canon is closed, and nobody can have experiences like the prophets and saints anymore. It comes partly for ethical reasons—the older Christian ascetic repulsion against hedonism and sensuality, the fear that drugs might encourage immoral activity. And it comes partly for theological reasons—the idea that God is transcendent and distant from the world, and could not be found in immanent experiences of worldly joy and natural insight. God is found only through suffering, self-denial, and striving, not through sensuality, which involves the world of the flesh and the Devil. Insight which comes through the world or through natural or man-made substances must then be evil, to be distrusted and banned. Only literal interpretation of sacred texts, approved sacred ritual, and rational theological argument is to be trusted.

Such theological perspectives have fueled a War on Drugs, in which the opportunities for ecstasy have been lost in the condemnations of evil, immorality, and insanity. And those drugs which can evoke ecstatic states under the right conditions have been lumped together with drugs that are addictive and even poisonous. Because ecstasy has been largely lost as a religious category, there can be no religious way to distinguish between them. Thus, if a person wishes to enter an ecstatic state by this method, he or she is forced into an underworld of crime, police, dishonesty, and danger. Is this where ecstasy belongs?

Another complicating problem here is modern society's appetite for entertainment. War and tragedy and sex are entertainment on film, fighting and competition are entertainment at arenas and stadiums, and religion is entertainment too. The sacred dances of Hawaiian hula and the Oriya devadasis are danced in nightclubs for entertainment, voodoo possession shows up in low-grade horror movies, even Christian sermons are judged on entertainment value as churches add bowling alleys, basketball courts, and movie screens. In the issue of ecstasy and drugs, this plays out in the original idealism and seriousness with which the drug exploration began, and the degeneration which occurred when the drugs became popular. People began to take them for recreational rather than therapeutic or spiritual reasons. The initial goals were obscured in the resulting fights between government control and individual rights.

In this subchapter, we shall explore how drugs are used for religious reasons, using as an example the religion of Santo Daime. It is based on the use of Ayahuasca, a tea made of two South American plants. It is their sacrament, taken in a ritual situation involving hymns, dancing and prayer, and its goal is to commune with Jesus as present in nature and the universe.

Santo Daime—Ayahuasca as Sacrament

Santo Daime is a modern religion, still in the process of formulation. It is a mixture of Brazilian spiritualism (generally Espiritismo, or more specifically Kardecismo[26]), African religion, Catholic Christianity, and shamanic animism, facilitated by the use of Ayahuasca (a drink made from the leaves of the plant *psychotria viridis* and the *banisteriopsis caapi* vine). In the Quechua language of Andean-Equatorial South America, *ayahuasca* roughly translates as "vine of the soul."[27] According to some translations, it is "the rope that links the world of the dead with the world of the living." In its older usage by the Amazonian Tukano Indians, Ayahuasca users are said to climb the vine to the Milky Way (the White Road or the Road of the Dead) to visit the ancestors. Such spirit flight often involved nature imagery, such as visions of jaguars and snakes.

In Santo Daime, this shamanic symbol is reinterpreted. The imagery of the vine is important, for theologically, Christ has returned to earth in the form of a vine and Santo Daime art often shows this vine wrapped around the Tree of Life. "I am the true vine" says Christ in the Book of John (John 15:1), and for Santo Daime, this vine is understood to reveal spiritual truth when it is ingested. Christ was metaphorically a vine in his physical body, and now he is physically a vine in his Second Coming, fully entering into nature. However, he is a vine who brings the person to heaven. Some interpretations say that Virgin Mary has also come back, in the form of the marijuana leaf.

The tradition was founded in the 1920s by a Brazilian named Raimundo Irineu Serra, later called Master Irineu. He was a border guard in the Amazon area of Brazil, and he went into Peru with a friend to explore tribal ceremonies. A vision that he had during one such ritual inspired him to found the religion of Santo Daime. He saw the image of a woman in the moon, and she said that she was the Queen of the Forest. She called him to go alone into the jungle to fast and pray, and drink a sacred tea. He was then to found a new spiritual path, the Santo or Holy Daime. She would be his guide and protectress as he began a religion based on spiritual journeys, using the sacramental drink called Daime.[28] She was later understood to be the Virgin Mary.

The religion was largely Catholic, though it incorporated elements from Spiritualist thought, such as karma, reincarnation, and belief in spirits. The name "Daime" came after the invocation in one of Serra's hymns, "Dai-me Amor, Dai-me Luz…" (Give me love, Give me light) and the religion came to be known as Santo Daime.

The Daime drink brings the *miracao*, the vision of the spirit realms. Jesus and Mary are here, but also other spiritual beings, who are perceived as tangibly present. The drink allows this spiritual vision because the Ayahuasca is itself the embodiment of a deity. It is a different incarnation of the Christ aspect of Jesus, in leaf and vine rather than flesh and blood. As the vine is made into a tea, an alchemical transformation occurs which is known as the *apuracao*. At this moment, it is declared that there is Mystery present, and the drink becomes a sacrament. It is then a gateway into the spiritual realm.[29]

This Christ consciousness, called in Santo Daime "Juramidam," entered the forest to "replant" the true Christian message. Such Christ consciousness begins as a seed in the heart, but grows into a strong and healthy vine of faith when nurtured by ecstatic religious experience. It emphasizes community love and support, and also helping nature—especially saving the rain forests of the Amazon.

By 1930, Irineu had a core group of followers, many interested in spiritual healing. He and his followers heard songs and hymns after drinking the consecrated Daime, and these were later sung for services. People wrote down the hymns, and started collecting them, as revelations or channeled songs from Juramidam. Hymns and songs (called *icaro*) became the major forms of teaching in the tradition, and many follow a call-and-response format. Ireneu's group in Rio Branco, Brazil, grew and spread. In 1981, he moved the center of the group to a village named Ceu de Mapia. Later followers revised the tradition, and Mestre Raimundo Sebastiao Mota de Melo, another important Daime figure, developed the ritual of using the drink during the mass or *hinario*. The Santo Daime liturgy consists of hymns and songs which are sung from dusk to dawn, and people dance together to them. It is important for people to be together during the ritual, which must be a shared experience.

The Santo Daime tradition emphasizes both vision and healing. Daime is medicine of the spirit, leading to spiritual rebirth, and the person is like an unborn chick breaking through the eggshell into authentic existence. This entrance into true existence comes through the *miracao*, a state of ecstasy which allows the return journey through the labyrinth of life, allowing the person to see and fix problems. As Alex Polari de Alverga states:

> We were in the miracao, the divine language and divine logic opening in our minds and felt by our hearts. Following the miracao is like walking on the blade of a knife: if we lose our concentration or courage in the middle of the crossing, we might experience vertigo… The bitter taste of the Daime seemed to adhere to my cells, and a strange energy was condensed

> within me. It was the force. The maraca's cadence contained ecstasy and joy within the limits of serenity... The force is a powerful energy that precedes the miracao. It is as though all the gifts, attributes, and vital functions of our body organs become visible... The miracao is the "third vision." The understanding that we acquire in it inspires a deep commitment to transformation... In the ceremony I saw all the people dancing in the rhythmic pattern as a living example of the oneness in multiplicity. A voice said, "Wake up now! Now you are truly a man, a Divine Being incarnated."[30]

This state occurred during a *hinario*, the ritual meeting with singing and dancing during which the drug is ingested. Alverga spoke of how the sacred plants allowed people to perceive that everything was within God, and human, animal, vegetable, and mineral kingdoms pulsated with the breath of life and were part of the eternal song of the Creator.[31] Images of evolution and transformation are an important part of Daime experience. As Alverga notes:

> With my eyes still closed, I saw the long journey of human beings... From a simple aggregation of cosmic dust, he or she became an organism able to receive cosmic intelligence and great evolutionary power. A being capable of asking about its own origins may remember the love that generated us. Thus we find the way back to the celestial dwelling where Christ himself has secured our destined Father for those who do not deny him. I saw all of this in a vertiginous form, like a movie in fast motion- wars, victories, sublime moments, and harsh punishments- the long journey of the human being on earth.[32]

According to Santo Daime, the goal of life on earth is to realize one's "superior self," the self which recognizes its union with God and gives knowledge of eternal life. There are various dimensions of the person: mental, emotional, astral, and spiritual. The drug is believed to bring these various aspects of the self to mind, so that the person may analyze them and fix problems. The astral aspect of the person includes "memories of the past" or past lives, when destructive beliefs and habit patterns were generated which still affect people today. It also includes ghosts and spirits, which may influence the mind and emotions. It is difficult to enter these deeper states; therefore, the Daime is needed. However, it must be used in a religious context in order to have the help of

Juramidam, who can guide the soul past its individual problems and into spiritual realizations. It is the state of *miracao* which reflects these breakthroughs. When the superior self-dominates the person, or is "seated on the throne," then all of the world is experienced as God, the world and the community are healed and regenerated, and the world is perceived as harmony, love, truth, and justice. Ecstasy is the process of penetrating these layers of the self and reaching the deepest layer of universal awareness.

God, or Christ consciousness, dwells in the *ayahuasca* vine as Juramidam, God within nature. Thus, it works not because it is a drug that affects the brain, but rather because it is a God incarnate in the form of a plant. When the person drinks the *ayahuasca*, the God enters the bloodstream and eventually the heart.

Daime ritual is organized and institutionalized. The rituals or "works" are the *hinario* or the mass, the "healing works" for spiritual healing, and the *feitios* or celebrations at the production of the sacrament. All of these rituals involve a strict diet, no drugs or alcohol, social hierarchy (a "godfather" or "commander" leads the ritual, aided by a group of "helpers" or "controllers" who act if there is disorder, and who help people to maintain proper posture and dance steps). People are grouped by age and gender, often in lines, and the hymns direct the voyages and help people who might get nauseated by the drug (this is a common side effect of the drug, which has given it the nickname "*la purga*"). These rituals are usually performed on Wednesdays, or on special days such as saint's days. They are long, lasting about twelve hours, with people dancing in specific formations and rhythms. Practitioners wear blue and white, the special colors of Santo Daime. The leader may be called the *curandero* or the *ayahuasqero*.

Ayahuasca has been used for centuries by the native inhabitants of the Western Amazon, and Santo Daime is not the only religion to use it as a sacramental drug. Today, it is also used in the religion of Uniao de Vegetal (a healing religion begun by Mestre Gabriel, who had previously been head of an Afro-Brazilian possession cult) and in the Barquinha ("little boat") group, which contains many elements of Umbanda spiritualism. Its traditional use has been by shamans for diagnosis and treatment of disease, divination, hunting, and warfare.[33]

It is part of the collection of species in the Amazon Basic known as *plantas maestras* or "teacher plants," which are believed by the local populations to facilitate entrance into the spiritual worlds.[34] Ayahuasca and other "teacher plants" reveal a living universe, in which nature is not only alive but conscious (though in a different way from human consciousness). By means of dance, prayer, song, and dream, people are able to become of mineral, plant, animal, ancestral, and divine forms of consciousness. The hymns which are being received by Santo Daime leaders and members include such sources as Biblical figures, saints, spirits of sacred plants, forest animals, other deities, and heavenly bodies.[35]

In his book *The Antipodes of the Mind*, Benny Shanon describes themes in ecstatic ayahuasca experiences. Ayahuasca can cause intense emotional experiences in the context of Santo Daime worship. He speaks of "a stupendous emanation of light, which was bountiful and full of love."[36] From the light emerged flowers, and "the entire world rejoiced in the wonder that is Life." There are visits to heaven which bring states of bliss and intense joy during flights of the soul. There is also great bliss during visions of Jesus and the Virgin Mary (who may appear as Queen of the Forest). Some cases involve intense joy and sorrow together, as in experiences of death and rebirth. Emotions may range from terror to exhilaration in a single set of visions.

Ayahuasca can cause a wide range of altered perceptions. Shanon notes visions of jaguars, ocelots and snakes, spirits, large trees, deities, and landscapes.[37] There are also visions of historical and mythological beings and scenes, the creation of the world and its evolution, celestial bodies and heavenly scenes, visions of light, experiences of death, and encounters with the divine.[38] Many of these involve panoramic historical visions. People may see events from the past and encounter animals and plants which appear to be conscious and aware. Many of these visionary creatures are from the Amazonian rain forest, where the vine grows. Sometimes there are serial visions with biographical scenes, in which the person relives the past and is able to resolve past conflicts. Some visions give ethical and moral guidelines for life.[39]

Ayahuasca can also change the person's sense of identity. People report union with God and sharing in the creation of the universe.[40] Nature is a part of God's glory, and people share in God's grace by being a part of nature and eating God's body in nature, the vine. Some Santo Daime practitioners identify with a universal life force or cosmic energy, which is their true identity. Such energy may be part of divine vitality, or it may be the *arutam*, the energies of the ancestors which uphold life on earth.[41] Such experiences are sometimes described by members of Santo Daime as entrance into eternity and a unity with all things.

Here, we see a religion whose sacrament is a drug, currently illegal in the USA, whose states of ecstasy have the elements that one would find in more traditional religious traditions, and whose ecstasies are comparable to the other forms of religious ecstasy in the world's mainstream religious traditions. However, Santo Daime has not yet developed a Theology which analyzes how ecstatic experiences occur. This is an area for future Daimista theologians.

Another area of "spiritual ecstasy" comes from the arts, especially music and dance. Here, we can examine the ways that these have come to be paths to ecstatic states.

Ecstasy and Rock and Roll—The Ecstatic Aspects of Modern Music and Dance

Music and dance have long been accepted as ways to enter religious and ecstatic states. In the modern West, however, they are not usually understood to induce these states. This is why responses to singers from Elvis Presley to the Beatles have been so strong—and often so negative. As Joost Meerlo describes a spontaneous outburst of rock and roll in a diner:

> From the very first note the young people near the small bar became untameable… A frenzied rhythmic seizure took possession of them: they yelled and shouted and rocked themselves more and more into a rhythmic trance until it had gone far beyond all the accepted versions of human dancing… A wild pantomime is displayed, resembling the rhythm of tribal utterances and the convulsive outcry of revivals… In modern Rock' n' Roll our youngsters dream and shriek and writhe and twist themselves as if wrestling with all the burden of civilized restraint.[42]

During the 1950s and 1960s, such music and dance in the USA were still associated with the primitive, the sexual, the unacceptable. However, in many other cultures, the music and dance that strongly affected people were understood to have religious importance.

In ancient Greece, the Harmony of the Spheres described the universe as a symphony of waves and rhythms, all forming together a sort of divine music. It was such music that kept the universe moving that caused the gods to dance in joy. Aristotle spoke of three categories of melody: ethical, which affected the soul; practical, which was linked with various occupations; and enthusiastic, which was associated with states of possession and ecstatic trance.[43] He also contrasted the Dorian mode, which he believed developed ethics and moral character in listeners, with the Phrygian mode, which was dynamic, passionate, and orgiastic. While the Dorian mode was calm and aristocratic, suitable for educating youth and building character, and "truly Greek," the Phrygian mode was emotional and dramatic, suitable for the lower strata of society and people in servile states, and basically foreign. Dorian lyre music was peaceful, inducing contemplative states, and Phrygian aulos music was passionate, inducing trance and possession—they differed like the north and south winds.[44]

According to one myth, the Dorian or Apollonian style of music originated with the God Hermes stretching strings across an empty turtle shell and hearing the cosmic harmonies of the universe. The Phrygian or Dionysian style developed from the cries of grief of the Gorgons after their sister Medusa had been slain.[45] In later medieval times, the Dorian mode became the basis of Gregorian chant, while the Phrygian mode developed into flamenco music.[46]

In indigenous cultures, music could cause people to enter into trance states either individually or as a group. One major form of trance music involved drumming, with the drum as the "voice of the spirits" for the shaman, and with prolonged drumming leading to concentration and then ecstatic states. Among the Tungus of Siberia, to play the drum was to have "mastered all powers."[47] The drum could bring order or frenzy and could be accompanied by chanting or by other instruments. Cross-culturally, some of the instruments most associated with trance states have been drum, flute, pan flute, aulos (double clarinet), lyre, lute, tambourine, fiddle, harp, gourd, and trumpet.[48] Some trance states are induced primarily through instrumental music: for instance, those of the Greek Corybantes, the Italian Tarantists, the members of Brazilian Candomble, and the trance dancers of Barong and Rangda listening to Balinese gamelan music.[49] However, we also see trances that use only vocal chants: in the Tunisian *dhikr*, the Wolof *samp* rituals, and the trances of the *bori* women of the Shango cult.[50] The major trance rhythms involve increasing tempo and volume (*accelerando crescendo*) and breaks or abrupt changes in rhythm.[51] Dance could bring both positive and negative ecstatic states. As Meerlo noted:

> Dancing man can be in tune with the universe, when his dancing will lead him to the pinnacle of ecstasy and unification, but he can also be out of tune, lost in the depths of confusion and bewilderment. In more sophisticated language we may say that dancing has an ecstatic, sacred action, or it can drag us into the netherworld of dance craze and chaotic feelings.[52]

Historically, there have been crazes of music and dance, where people were drawn from their homes into the streets, as in the story of the Pied Piper. There have been contagious dance epidemics, especially in fourteenth- and fifteenth-century Europe, with people imitating animal howling or other behaviors. The most famous were probably the St. Vitus Dance epidemics in the Middle Ages and the outbreaks of Tarantism in Southern Europe. Later some of these became stylized as protective or expressive dances.[53] There were ecstatic dances around statues, fire dances over burning charcoal, jumping dances as protection against possession, and processions and death dances. Meerlo notes that dancing leads to a group identity:

> The religious dance leads to the ecstatic feeling of mass participation, of being merged with and absorbed into the greater dimension of mass-elation. The Greeks spoke of the Dionysic frenzy… The mass-suggestion of common religious ecstasy was often used in a therapeutic dance to cure disease. Our modern dance is full of remnants of these ancient ritualistic experiences.[54]

Such dances have often been feared as leading to social chaos. And one approach that societies have often used to maintain social order has been the discouraging or banning of ecstatic religious or dance rituals. However, secular dance has also been a way to enter ecstatic states. As Iris J. Stewart states:

> Dance by its very nature is ecstatic. Almost every dancer who has written about dance makes reference to ecstasy, although she or he will often describe the experience in other terms. The experience of ecstasy is associated with a feeling of timelessness or eternity, adding another dimension to our familiar temporal existence. One may have a sensation of elevation or levitation, the feeling of a rising spring or a flowing stream. There may also be a sensation of heightened inner consciousness or awareness, or an increased feeling of stillness and peace. Freedom from self, from guilt, from sorrow and desire accompanies ecstasy.[55]

As Stewart defines ecstasy:

> Ecstasy is a feeling of glory, joy, happiness, satisfaction. It is the state of mind that ensues, however briefly, from a religious experience, when everyday consciousness is stripped away, leaving only the essential self. Ecstasy creates both a catharsis and a creative inspiration, making the vessel of the body empty and fit for the divine to enter.[56]

While much of modern dance emphasizes performance, ecstatic dance needs neither onlooker nor witness—it evokes an inward state of transcendence, based on rhythms, breathing, and physical movement. While it is an activity that evokes this state, it is paradoxically a state of relaxation:

> What is commonly called the ecstatic state is really a deepened state of relaxation. You are in control of your behavior; you are lucid, conscious, and able to concentrate, and can 'awaken' at any time. Ecstasy is not easily or deliberately acquired, but if you watch for it, it may come to you more easily than you think.[57]

One major approach to modern ecstasy via music and dance is the rave (called Dance Music Culture or Club Culture by participants). Raves are usually said to have begun on the Spanish island of Ibiza in the mid-1980s, where vacationing hippies, musicians and artists, and celebrities would dance all night to trance-inducing music. They called this music "Acid House" and the island became a haven for Northern European tourists. By 1986, Ibiza came to be called "XTC island," after the use of the drug MDMA or Ecstasy that was used to accompany the all-night dancing. From Ibiza, raves spread through Europe and the USA. Though their numbers have declined in recent decades in the USA, they are still popular in Europe.

Raves began in England during the early 1990s, partly as a response to the social stratification and competitiveness of British working-class life. They used a form of music or "techno sound" with a repeating rhythmic pattern of about 125 beats per minute (double the normal heartbeat). The music is generally played at high volume, accompanied by otherworldly electronic effects.[58] Raves became very popular, with somewhere between half and million and two million people in Britain alone going to raves every weekend. At the Castlemorton rave in 1991, there were 25,000 people in the English countryside, dancing to hypnotic beats and tribal images.[59]

Music has long been a way of creating countercultures in the modern West—jazz and blues in the 1920s, rock and roll in the 1950s, hippies and the Grateful Dead in the 1960s and 1970s, 1980s reggae, and 1990s raving and dance music culture. All of these have continued into the second millennium. However, the earlier forms were more sensual and less spiritual, rebelling against conservative social and sexual values. The later forms were influenced by ideas of generic spirituality, tribal community, and a return to nature. They sought to bring ritual to people alienated from religion in an institutional setting and to generate joy and mutual trust.

As one writer/participant describes it, raves or DMC (dance music culture) are based on "Technopaganism," which mixes old tribal traditions with contemporary technology and brings a feminine style of ecstatic experience to people who rebel against the aggressive style of the non-ecstatic dominant male culture.[60] Raves try to counter the hierarchical aspects of modern society with a new form of egalitarian and tribal community, and communal ecstasy is its rite of initiation. As one group, Return to the Source, describes it:

> The all-night dance ritual is a memory that runs deep within us all; a memory that takes us back to a time when people had respect for our great mother earth and each other. Dancing was our rite of passage, our shamanic journey into altered states of reality where we embodies the Great Spirit and the magic of life... The ancient memory has reawakened, the all-night dance ritual has returned. At Return to the Source, it is our vision to bring back the dance ritual. A ritual is a sacred act with focused intention. We aim to create a modern day positive space, created with love where we can join as one tribe to journey deep into trance, just as our ancestors did.[61]

There are a variety of techniques used by ravers to change their traditional ways of understanding the world. They attempt to alter biological rhythms by sleeping in the day and dancing at night, opposing the "overculture" by "phase-locking," or harmonizing with the internal rhythms which run through humanity and the cosmos.[62] The use of the drug Ecstasy is understood to change the heart, making people more empathic and open, creating love and harmony, and a common sense of identity. The group identity is understood to challenge the "over-culture," which is obsessed with sexual conquest and money. And the current generation of "screenagers," who are immersed in technology (especially via the Internet), understand themselves to use art and technology to live in a new kind of world, beyond the limits of space and time.[63] Thus, there is all-night dancing to alter the body, new technology to alter the mind, and Ecstasy to alter the heart. As a computer engineer describes his experiences:

> Looking at the joy others around me are experiencing causes me to dance harder, which in turn forces them to dance harder, and the feedback loop continues, driving us further into that state of transcendental bliss. Predawn, the universe is in complete order, everything has worked out... It is as if Mother Nature, or Gaia, or God, or whatever you wish to call the spirit that encircles this planet- with all of its infinite duties of spinning the earth around the sun, keeping balance between a trillion living organisms- has stopped, and paused, for just 60 seconds, and acknowledged me as a life form. For the first 60 seconds of the new day, Gaia has decided to show me, one mere human, all of her beauty and glory in one concentrated burst. I am floating above the earth below me, experiencing the most sheer beauty I have ever imagined experiencing in my life. Everything around me is pure light, pure positive energy, pure love and pure beauty all at once... I am being touched by the hand of God.[64]

Simon Reynolds describes a rave as a situation in which the "music-drugs-technology nexus" is fused with spirituality and "vague hippy-punk-anarcho politics" to form a 90's counterculture. Because of its origin on the tourist island of Ibiza, the rave has also been described as "internal tourism," a holiday from ordinary life.[65]

Different types of raves developed over time. Some emphasized spirituality, with DJs as modern shamans leading vision quests. In this Technopagan approach, they would work crowds into ecstatic frenzies by playing strange and original forms of music. They created "white labels," unmarked and unnamed tracks, and bought rare promos, remixes, imports, dubplates, and limited release cuts. The DJ (or Technoshaman) would control the emotions of thousands of people, "leading a cosmic collective and spiritual journey."[66] They might mix the sounds of African and Native American tribal drums with the sounds of video games and traffic jams. Music was cross-cultural, full of intertwined sounds and rhythms. It was clipped, spliced, layered, tweaked, lengthened, shortened, compressed, and digitally remastered. The goal was to create alternative ritual sound, and for the devotees and participants to enter spiritual realms and ecstatic states based on these sounds. By doing this, rave kids sought to create a better planet and a new unified culture, less aggressive and competitive than the culture into which they had been born.[67]

However, there were also rave groups that were more recreational, with people simply trying to escape the boredom of working life. For these groups, there was little talk of spirituality, and more of social networks and communities of friends. These were not ecstatic groups, and the focus of the rave was entertainment.

Several types of electronic music developed from the raves. Club music was complex and drug-oriented, and included such types as house, acid house, and garage music. Hardcore music was very fast and break beat and was mostly found at teenage raves—in such forms as jungle, gabba, and trance music. There was also music for people who liked the trance state but tired of the rave lifestyle.[68] These forms of music became more complex, more "sampladelic" using such digital technology as the sampler, a computer that could convert sounds into digital code, from which it could be easily rearranged into modern "hallucinogenres."[69] Music moved from "naturalistic" (recording a band in performance) to "digital psychedelia" (creating new forms of music intended to generate new states of consciousness).

Such music grew more popular as raves grew larger—during the 1990s, such British megaraves as Amnesia, Raindance, Fantazia, and Dreamscape drew crowds ranging from 10,000 to 25,000 to dance all night in deserted hangars or under circus-size tents.[70] In the USA, there would be smaller crowds of up to 7,000 people, who would meet in unused warehouses or rural areas. Part of the drama was finding the raves in the first place—they were not publicized, and tracking them down involved finding flyers, hearing voice mail messages, and getting directions to their secret and ever-changing locations.[71] While there was little ecstatic dimension to the recreational raves, there was a strong ecstatic emphasis in the spiritual ones. As T. Scott notes, modern religion focuses on experience, not dogma:

> Music is utilized as a vehicle for discovering religious ecstasy: music never makes any claims for salvation or repentance but only [acts] as an auditory

> guide on that ecstatic journey. And counterculturally, youth culture correlates music and the use of drugs in defining its new religious style.[72]

While much of this ecstasy came through music and dance, some also came through Ecstasy (the drug MDMA). As sociologist Mary Anna Wright notes,

> I have met hundreds of people within the dance scene who had had profound experiences caused by a combination of drugs, music and communal euphoria. Such ecstatic experiences may not be intentional but always have a deep impact on those who share them. Interviews with people involved in all genres of dance music have revealed what I believe to be a pharmacological shaping of culture. This means interactions within dance culture are largely based around the effects of MDMA.[73]

Popular culture has greatly distorted both the uses and effects of Ecstasy. This is partly because of the general drug war of the USA, whose proponents randomly lump together all psychedelic drugs as equally evil and dangerous, and partly because there are many pills which are called Ecstasy, including ones which bear no relation to MDMA. There is thus wide confusion over the identity of Ecstasy. The author Simon Reynolds, publishing in 1998, estimated a one in ten chance of buying Ecstasy that had no MDMA at all (instead containing decongestants, antihistamines, or other substances), and a 66% chance of buying adulterated MDMA.[74] The odds since then have gone up considerably on getting Ecstasy with no MDMA, according to other writers.

MDMA was first developed and patented by the Merck Pharmaceutical Company in Germany in 1912, shortly before World War I. According to one story, it was to be used as a weight-reduction pill. According to another, it was intended to be used as an appetite suppressant for German troops. However, with its side effect of generating empathy, it would not have been useful in a combat situation. Reynolds notes that when it was later used in the 1990s in experimental therapy sessions for soldiers from Nicaragua who had been traumatized, 75% of the subjects expressed a desire for peace and an end to war. Some spoke of loving everybody, even the enemy.[75]

MDMA was rediscovered in the 1960s by Alexander Shulgin, a biochemist working at Dow Chemical. He distributed the drug to psychotherapists who were interested in its ability to aid the psychotherapy process, especially in helping patients work through difficult emotional problems. It was also found to be beneficial in marriage therapy and in dealing with fearful situations. As one counselor and writer noted about her MDMA experience:

> I was always afraid of death and now I'm not afraid of death. Because I tasted what death is- no body. I was all mind, in communication with every mind that ever lived. And that was an *extraordinary* experience and I'll never forget it. This was my number one gift of all my life. I'll never have anything as wonderful as this... for me it was the most profound experience of my life, absolutely.[76]

The same writer witnessed husband's death from cancer and supports the use of MDMA for terminal illness:

> This drug can allow people to die well. They die well in the emotional arms of their families. They get reconciliation, they talk about things they normally would never talk about... I think it's so sad that we don't have this in every hospital, not for just cancer patients...[77]

As another older MDMA user stated (at 72 years old):

> It is empathizing. It is so reassuring and rewarding- allowing a renewed access to feeling which has been pretty much eroded out of our culture. Too much thinking, not enough access to feeling. And the feelings are so warm and so trusting that I think it's a precious gift and we're not learning how to use it.[78]

While modern popular culture emphasizes teenagers using the drug, the use of drugs by psychotherapists has primarily been with adults, including the elderly.

The drug's early supporters wanted to call it an "empathogen," a drug which enhances feeling, and it was used in the 1980s for therapeutic purposes. It was shown to motivate compassion and hope, for the stress situations of the patients and toward the world at large. However, word of its effects spread, and people began to use it outside of the therapeutic setting. It became popular as a recreational drug, and promoters started to distribute it in nightclubs and at parties. There was less concern with the purity of the drug and much concern with making money by selling it—often in adulterated form.

By the 1980s, the DEA had determined that Ecstasy should be banned and classified as a Schedule I drug (a drug with a high potential for abuse, no medical use, and lack of safety under medical supervision). The DEA called hearings to examine the uses of MDMA, and there were numerous witnesses. Despite minimal evidence of adverse effects, and the DEA administrative judge's ruling that the drug should not be banned (it should be categorized as Schedule III, to be legally used only by medical professionals), the DEA decided to ban it anyway.[79] Ignoring their own judge, and basing their decision on a single set of experiments (which have since been disproved), in 1986 the DEA made MDMA an illegal drug.

This was a great blow to the psychoanalysts who had been using MDMA in a therapeutic context. But it brought Ecstasy fully into the counterculture. Rather than the therapeutic ecstasy of having patients being released from pain and trauma, the focus shifted to rediscovering the shamanic ecstasy of Native American and Aboriginal cultures, to developing ecstatic music and dance as part of a spiritual life, and even to influencing the social climate (creating peaceful interactions in Northern Ireland, and at football games between rival factions, as examples).[80] It came to be used for group communication and for spiritual practice. As a Benedictine monk noted, Ecstasy creates intimacy which is not erotic but rather philosophical and mystical, and a sense of "inner communication" which is rarely achieved in ordinary discourse.[81]

Raves have been moving in new directions. Rave style has been adopted by several religious groups interested in having younger members. Thus, we have "Torah raves" (with Kabbalistic practitioners who dance out the Hebrew letters) and "Christian raves" (with the tribal community of Jesus, and advertising flyers that show teenagers in states of bliss). There are also "Yoga raves," where people who were once "club kids" go to all-night parties involving yoga and meditation. In Chicago, the TranceZenDance Tribe also has "meditation raves" whose goal is "linking consciousness with everyone on the planet," while Body Temple is a "tantric circus" in which people perform rave-like shamanic trance dances to find their inner "power animals."[82] These are raves which do not use the drug Ecstasy, but share the ecstatic goals of universal consciousness, joy, attaining trance states, and the realization of tribal or group identities. We also see the emergence of a conscious "rave spirituality." This is reflected in religious discourse about clubbing and clubs with names like "God's Kitchen" and the "House of God," as well as club flyers which make use of religious iconography.[83] We see the religious language used more clearly in the development of trance dance.

Trance Dance, Neo-shamanism, and Universal Peace

Not all dancing and drumming into ecstatic states involve raves and large concerts. Alternative approaches to shamanism have developed out of New Age sources interested in inducing ecstatic states without the use of modern drug and music technology. Indeed, these approaches are more interested in traditional cultures and their techniques of return to nature.

Trance Dance has developed as a ritual practice based on neo-shamanism. From this approach, the cultural specificity of the various forms of shamanism is ignored, in favor of general themes which are understood to run through the various shamanic cultures and tribal groups. Thus, we have people interested in attaining trance states, who are willing to do dances from many cultures and belief systems: Native American dances, possession dances of the Yoruba, Umbanda and Candomble traditions, masked Balinese and North African dances, tai chi chuan. The goal is to attain awareness of Spirit in whatever cultural form it manifests.

Neo-shamanism is a New Age appropriation of aspects of shamanism from various different tribal groups, using rituals and sometimes deities as people create their own spiritual paths. The meaning of the term "shamanism" has shifted over time, and now it most broadly refers to any non-institutional set of religious beliefs and practices, which are ideally individual, original, and creative.

One influential book for this approach was Michael Harner's *The Way of the Shaman*, which was basically a popular handbook on how to become a shaman without doing fieldwork, getting apprenticed to a shaman, or indeed learning about any shamanic cultures. Harner distinguished between the ordinary state of consciousness (OSC) and the shamanic state of consciousness (SSC), comparing them to Carlos Castaneda's "ordinary" and "non-ordinary" realities.[84] Using some of the themes from Mircea Eliade's famous and more academic book *Shamanism: Archaic Techniques of Ecstasy*, Harner developed techniques for bringing people on shamanic journeys. These involved drumming or cassettes of drumming, and visualizations. The drumming should be 205–220 beats per minute to be most effective, with more rapid drumming signifying the end of the journey.[85] There may also be "power songs" which involve chanting. Visualizations include going down into an opening in the earth, entering a hollow tree, and finding a tutelary spirit or guardian (often in animal form). Harner's shamanic rituals may also include dancing, with masks or costumes, imitating the cries of animals, and having dancers assume the personalities of animals.[86]

This idea that it is the music and dance itself which induce religious or spiritual experiences, rather than the specific tribal knowledge, identity, and history, became the basis for the development of a variety of new forms of dancing and drumming rituals. Such experiments have been going on since the 1960s, as an exploration of how traditions from other cultures might be relevant to the modern West. One set of dances came to be known as Trance Dance.

There are a variety of practitioners of forms of Trance Dance. Frank Natale, a devotee of the late guru Muktananda, developed one of the earlier ones in the 1960s. As he stated:

> Through ancient breath patterns, movement and rhythms which impact the brain, the Trance Dancers begin to see with their senses rather than their eyes. Other times and other forms reveal themselves, and the Trance Dancer's consciousness is filled with the magical world of Spirit… While your body dances, your soul travels and remembers, free of the limits of this life, free of the limitation of time and space.[87]

The practices he emphasizes in his form of Trance Dance include soul hunting, masking and use of ceremonial costumes, the use of power objects (such as crystals, amulets, rings, and photographs), drumming, the use of teacher plants (or entheogens, plant-based psychedelics used for religious purposes), controlled breathing, and initiation rituals. Soul hunting is based on the idea that souls can be "lost" or broken through trauma—people in some sense lose pieces of themselves through suffering, and these lost aspects must be recovered in order for the person to achieve spiritual or psychological wholeness. In the trance state, the person can go back into the past, and find these lost parts of the self and reintegrate them.

The philosophical background of most shamanism is animism—the belief that humans, animals, plants, and minerals all have consciousness, as do ancestors and nature spirits. The original deity is the Universal Spirit, the "incomprehensible collective consciousness." Realization of this Spirit causes intense ecstatic joy, even a "conscious collective orgasm."[88] It may be personified in various ways, and the most important one is as a goddess, often called Great Mother Earth. One way to realize this Spirit is to dance out or express one's Spirit Animal. This imitation of animal movements and sounds is believed to connect the person with his or her totem ancestor, and to give insight into life's problems through discovery of a deeper identity. People connect to the Spirit Animal, and to nature itself, through rhythm:

> Our life vibrates with rhythms as we quiver to our mother's pulse in her womb until our dying heart's final vibration at death. These and other rhythms in our life connect us to the timeless rhythms of day and night, new and full moon, seasons and stars.[89]

As Amoda, another teacher of Trance Dance, notes, there is a great emphasis on dance itself as a ritual action.

> To dance is to listen to the rhythm of your heartbeat and follow the pulse of the deep within your bones, to flow with the current of your feelings and let yourself be moved, to surrender to the passion of the moment and take flight into freedom. To dance is to resonate with the core of your being, vibrate with the hum of creation and express the uniqueness of your soul. To dance is to be fully alive, fully awake and filled with Spirit... to dance is to be ecstatic![90]

Such dance allows for the expression of intense emotions and for identification with unusual states of being and mythical beings. Sometimes it uses rituals and symbols from particular traditions, such as Kundalini yoga imagery. Here is what one dancer says of her first trance experience, which has little of the shamanic symbolism, but clear Kundalini imagery:

> I am flowing, spinning, beyond time and space... I do not know who or where I am. I am lost in a kaleidoscope of shapes, colours and sounds... A rush of energy like electricity shoots up my spine and right into my brain, my head explodes into white light and I am showered in sparks coming out of the top of my head. I stand transfixed, rooted to the spot, my head tilted back, my whole body trembling with ecstasy. I am bathed in love and light, the most sublime orgasm I have ever had![91]

Another dancer seems more familiar with the shamanic origins of the practice:

> Beating drums and beating heart… I descend into darkness… I am in a cave and I am wild, ancient and free. I am a bear searching in the dark, sniffing and sensing so many things… now I am a barbarian slave dancing naked for the Tribal Chief, a dance of possession, seduction, fear and exhilaration. I fade into nothingness, engulfed by pure blackness and then blinding light… Now I am a huge bird soaring over mountains and forests… I see so far, so clearly. I can hear the drums in the distance, they are getting closer now… I am back in my body as I know it, back in this large hall filled with so many people… a tremendous peace fills me.[92]

Modern trance dance has also become associated with yoga practice, as seen in the trance dance practitioner Shiva Rea. She focuses on Vinyasa or "breath-synchronized movement," and is known as a teacher of Prana Flow Yoga and Yoga Trance Dance for Life. She was trained in the Krishnamacharya lineage of yoga as well as studying Tantra, Ayurveda, Bhakti, Kalaripayattu, and world dance. She was involved in a 2011 debate with traditionalists at the International Yoga Festival in Rishikesh (the world's largest yoga festival) over how much Indian influence should be present in yoga practice.[93] She felt that wearing tight, low-cut clothing could be suitable for yogic meditation and trance dance, while they felt that modesty was more appropriate. She posed for Vanity Fair magazine as "The Madonna of Yoga" doing a yoga pose topless with tigers nearby, as a way of emphasizing the freedom of yoga and dance for both body and mind.

Trance dance gives the imagination freedom to roam, but it may also be guided by rhythmic drumbeats. Drumming and dancing have partly been popularized by Michael Harner's emphasis on the importance of drumming for shamanic experience and partly by the interest in modern indigenous religions and their rituals, especially Native American and Australian Aboriginal religions. Drums are said to have a voice, to give knowledge, to speak if the person is willing to listen. Michael Drake states that the drum attunes the person to ecstasy, like tuning a radio to the desired frequency, and that drumbeats can evoke universal consciousness:

> Everything in the universe, from the smallest subatomic particle to the largest star, vibrates with rhythmic motion. Rhythm is the heartbeat of life. The drum's beat unites the shaman with all life forms into a single being, a single heartbeat… Drum rhythms cover the whole range of human feeling.[94]

He notes that drum knowledge may come through visions, symbols, images, or intuitions, and the drum carries prayers and intentions upward. The shaman must "orchestrate the inner imagery," to contact spirit guides and power animals and to see past, present, and future events.[95] He or she may also communicate with plants and enter the tunnel to the lower worlds to encounter the spirit form of the plant (in human, animal, or mythical form). Drumming can make the shaman become a conduit for the energy of the stars, put a person in harmony with the celestial bodies, and make him or her or aware of the vast network of energy that is the spiritual form of the earth.[96] Drumming allows transcendence of time and space, by guiding the person on spiritual journeys. An "energetic relationship" to the drum is required, which involves giving the drum dedications, blessings, and offerings, calling in goodness, holiness and power, and inviting spirit guides and power animals to help out.[97] Only when the drum is empowered can it generate ecstatic states.

When the drum is the voice of the spirits, many things are required. Here in Charleston, we sometimes hear of drums being "baptized" in order to speak the words of God. Such rituals are associated with the Sea Island tradition of Gullah religion, and also with Haitian Voodoo, where many deities may come down to bless and possess objects. When dancers hear the voices of the gods through the drum, they may get possessed during the dance, ridden like a horse, and the gods will express their intentions through the dancers.

There are a variety of modern "dancing and drumming" and "trance dance" traditions which have been developed since the rise of the New Age that see themselves as loosely based on older shamanic traditions. We see Biodanza, Barefoot Boogie, Leo Rutherford's Shamanic Dance, Osho's dancing Nataraj meditation, raves and their "rhythm doctors."

Another ecstatic tradition of dance was developed by Gabrielle Roth, who refers to herself as an "urban shaman." In her book *Maps to Ecstasy: Teachings of an Urban Shaman*, Roth uses music and dance as a spiritual path. For her, ecstasy is both a path and a goal:

> Ecstasy is my high, a natural state of pure being. Ecstasy is my experience of god. It is a state of total aliveness and unity, unity of body, heart, mind soul and spirit. It is what we need to heal our psychic dismemberment.[98]

Ecstasy involves trance, and Roth was subject to trance states from her childhood: "I easily slipped into trance, could see through bodies, smell death, feel birth, know when someone was in pain and how to guide them through it."[99] She scorns such "cheap imitations" of ecstasy as pleasantness and sentimentality (she calls these the degraded forms of joy and compassion), and the other imitations that people seek:

> For obviously, ecstasy isn't simply infatuation, titillation, some temporary high that comes through drugs, ephemeral achievement, etc. Such things are simply intimations, if not imitations, of ecstasy. Rather ecstasy means being fully alive.[100]

Her path to ecstasy is dance. In dance, the body, the emotions, and the heart are freed of traumatic blockages, for people need expression of emotion as athletes need exercise. Roth states that blocked emotions stunt growth and behavior, cause emotional inertia, pervasive numbness, waking death, and a tendency toward disease.[101] All of these are blocks to intense experience.

Roth's path to ecstasy through music and dance goes through five layers of consciousness: inertia (where the person is frightened and passive), imitation (where the person follows society's ideals), intuition (the chaotic beginning of personal creativity, often involving disillusionment with empty traditions), imagination (spontaneity harmoniously mixed with skill), and inspiration (transcendence of conscious effort, an ecstatic level of acting and being, suffused with energy and the spirit of life).[102] At the highest level of inspiration, the person can reshape the culture through new myths and rituals.

These states are associated with five basic emotions (fear, anger, sadness, joy, and compassion) and five basic rhythms (flowing, staccato, chaos, lyrical, and stillness).[103] The urban shaman is a practitioner who lives in the modern world but recognizes the value of intense and archaic states. Shamanic ecstasy (which here emphasizes intense emotion) is a state of spiritual awareness which does not involve a deity or belief system or religious institution, but simply emphasizes ritual behavior as salvific in itself. It recognizes an alternative tribalism of people with different values, especially in terms of what makes the good or meaningful life.

As Amoda notes, for many people it is hard to have a broad identity like "world citizen." People require smaller group identities, as a basis of friendship and cooperation. Thus, we have the modern-day tribes, whose values of bonding, loyalty, justice, and honor act as a basis of community and work against the stress, violence, and ecological disasters of the modern world. Such harmony and cohesiveness work against the current breakdown of family and community. She notes that there are other alternative groups besides spiritual ones—there are sports fans, rockers, punks, hippies, ravers, jocks, mall rats, groups defined by their music, clothing, friends, and places to spend time. While some groups are more violent—such as gangs and some religious fundamentalists—most try to create community in more peaceful ways. A good example for her is the Rainbow Family tribe.[104] This "new tribalism" has become the basis of many alternative spiritual traditions.

But we return, in the end, to rock and roll. There are many descriptions of the passionate, overwhelming responses of fans to singers and musicians like Elvis Presley, Eric Clapton, and Madonna and to singing groups like the Beatles, Rolling Stones, Grateful Dead, and U2. As noted earlier, Gary Laderman understands such emotional responses in the audience as serving a basically religious function. But we also see ecstatic states in the musicians themselves, both in composing music and in playing it.

Another well-known musical group is the rock-jazz fusion band, the Mahavishnu Orchestra, which uses ecstatic and religious imagery in its music. With albums titled "Between Nothingness and Eternity," "Inner Worlds," "Passion, Grace and Fire," and "The Inner Mounting Flame," there was clearly an interest in spiritual imagery. As the music critic John W. Patterson writes of its leader, John McLaughlin:

> John McLaughlin's music went beyond mere music, beyond jazz, beyond rock – it housed a soul, it reached into spirit and the visions within all became new. Sure enough, McLaughlin knew jazz, rock, Eastern Indian music, and melded it all into a powerhouse of jams that blew most everyone away in the jazz and rock worlds. But The Mahavishnu Orchestra forged more than music – they delivered a *religious* experience. Things McLaughlin needed to say, were expressed through sound, words were spoken beyond hearing, echoes of a vital transformation filled each composition. *Birds of Fire* was one of my first experiences in hearing the "fire of the soul" coming through the medium of music… I heard bliss, frustration, anger, anticipation, elation, fury, ecstasy, euphoria, sorrow, joy, power, imagination, dreams, hope, stress, release, passion, and so much more. It's all there – and if you cannot feel it when you listen – you have missed the rawest power of The Mahavishnu Orchestra and you therefore have yet to really "feel the tingle" up your spine, the strange rush of winds down the "halls of your soul."[105]

Musician magazine, in a 1993 survey of "The 100 Great Guitarists of the 20th Century," referred to John McLaughlin as "the mystic credited with inventing real fusion." As McLaughlin stated in an interview with *Down Beat* magazine, "My work in music is a work of the spirit, it's a development of my spirit, and the development of myself as a human being." McLaughlin became a disciple of the Hindu guru Sri Chinmoy and took the initiatory name Mahavishnu. He sought in his music to unite Eastern and Western forms of music, and he has said that his spiritual development over the past fifteen years has probably been the most significant part of his life. As he states in a 2010 interview:

> "In the end, this is really all that we have," McLaughlin points out. "How do we live with our soul? How do we live with the supreme being? What kind of relation do we have with ourselves through our soul?"… "This is what's great about jazz music or improvised music," he adds. "You're telling the story right now in its realness. What's weird is that you can go to the stage tired — you didn't get much sleep, you haven't eaten, you have to get up early, you had three flights… But you get on stage and you get amazing inspiration. You don't know where it comes from. And that's all, really, that we live for when we go to the stage — to have that thing happen. Because that's the only moment we have, really. The rest doesn't exist."[106]

The rock musician Carlos Santana admired the Mahavishnu Orchestra and John McLaughlin. Aware of Santana's interest in meditation, McLaughlin introduced Santana to his guru, Sri Chinmoy, in 1973. Santana was given the name "Devadip," meaning "The lamp, light and eye of God" and he recorded an album with McLaughlin, called "Love, Devotion, Surrender." Santana later claimed inspiration from the kabbalistic angel Metatron, who prophesied the future success of his album "Supernatural," and he described his relationship to the angel in *Rolling Stone* magazine.[107] Santana also claimed supernatural inspiration for his album "Supernatural," including communication with the spirits of Miles Davis and Jimi Hendrix. He noted that Metatron wanted him to help listeners realize that they themselves have a deeper divine aspect:

> Metatron wants something from me, and I know exactly what it is… The people who listen to the music are connected to a higher form of themselves. That's why I get a lot of joy from this CD, because it's a personal invitation from me to people: Remember your divinity.[108]

Metatron is an archangel of light described in some forms of the Jewish mystical tradition of Kabbalah. We also see kabbalistic imagery in the rock-reggae music of the singer Matisyahu, who studied with the Chabad Lubavitch Hasidic group and in 2006 was named Billboard's top reggae artist. He writes of his yearning for God or Hashem (rather than the reggae term for God, Jah) who is like "water for my soul" when it is thirsty, and he gives himself fully to God from "the essence of my being." Indeed, "Hashem's rays fire blaze light my way light my life… Reaching for my G_d like skyscrapers in the night." Matisyahu later encountered a Hasidic group, the Pinsk-Karlin group in Mea Shearim, Israel, where the Hasids would scream their prayers, "not just at moments of peak religious ecstasy, but throughout the service." He liked this and left the Lubavitch group and aligned himself with the Karlin group in 2010.[109] Commentators note that he "brings hip-hop and rap to the religious ecstasy of the Hasidim."

There are also many varieties of Christian rock music, whose singers describe the presence of the Holy Spirit, speaking in tongues, conversion and being "born again," visions of paradise, and the ways that Jesus is living in their hearts. An example of a popular Christian rock group is Petra (it has won three Grammy awards and nine Dove awards). Their lead singer is John Schlitt, who notes that God speaks through Petra, and that their eyes are continually fixated on Jesus.[110] While Christian rock is perhaps the most mainstream of the spiritualized rock and roll approaches, it is often attacked by more conservative forms of Christianity. It has been portrayed as the Devil sneaking "demonic jungle music" into the church, helped variously by hippies, the Illuminati, and vampires. One major concern of its opponents is whether it is merely Christianizing pagan ideas and symbols, and stirring the flesh rather than the spirit.[111]

All three of these problematic areas—sex, drugs, and rock and roll—were often mixed in the songs of some of the bands of the 1960s and 1970s, whose music was inspired by LSD and other psychedelic drugs. One group, The Doors, took their name from Aldous Huxley's book *The Doors of Perception,* which described some of his experiences of cosmic union after taking mescaline. Psychedelic rock bands tried to induce this sort of state, to help people "break on through to the other side" by combining rock music and imagery of psychedelic drugs. These included the Grateful Dead, the Incredible String Band, the Velvet Underground, Pink Floyd, The Moody Blues, H. P Lovecraft, and

King Crimson, among others. Songs inspired by altered states of consciousness included the Byrds' "Eight Miles High," Jefferson Airplane's "White Rabbit," Pink Floyd's "Interstellar Overdrive," and Quicksilver Messenger Service's "Fresh Air (Have Another Hit)." These songs were often understood to refer to ecstatic and mystical states, like the "cosmic consciousness" or *brahman* found in Hindu Vedanta.

Such music was often combined with meditation, and ex-psychology professor and drug promoter Timothy Leary developed the new profession of psychedelic trip guide, a person who would have "the patience of a first-grade teacher, the humility and wisdom of a Hindu guru, the loving dedication of a minister, the sensitivity of a poet, and the imagination of a science fiction writer."[112] He viewed psychedelic drugs as having the potential to lead to states of transcendent consciousness with correct guidance, and Leary's International Foundation for Internal Freedom supported the use of drugs (along with sexuality and music) as ways to attain ecstatic states.

The issue of whether such altered states could be considered as equivalent to traditional mystical and ecstatic experiences has been an ongoing debate. While the mixture of rock music and psychedelic drugs with sexual freedom could create intense inner states, it also created fear and resistance in outside observers, and greater legal and cultural limits on all three areas. These experiences have been viewed as threatening to more conservative religious groups, who define acceptable religious experiences much more narrowly and believe in following a revealed plan for human life; these forms of ecstasy violate moral law and are considered to be contrary to the divine plan of salvation.

But the range of allowable mystical and ecstatic states in mainstream religious traditions can be considered to be so narrow as to be virtually invisible. As Victoria Nelson has noted, "the expulsion of religious experience itself from the main currents of Western intellectual culture" has forced the transcendental into the "low art" of secular entertainment.[113] This was necessary to fill the vacuum left by the loss of religious ecstasy in the Theological and academic spheres. Thus, ecstasy has gained a special place in popular music and the arts. Nelson notes that forcing religious experience into the aesthetic realm anesthetizes it, making it imaginary and unreal. Today, people can only appreciate such experiences if they do not really believe in them. Thus, the spiritual and the transcendental can only be experienced if they are denied and displaced into the arts and popular culture.

In this chapter, we can see how sex, drugs, and rock and roll can be reinterpreted and developed as rituals intended to induce spiritual states and evoke ecstatic states and religious goals. They can be used for pleasure, or for love, or for religious, personal and cultural insight. While their hedonistic aspects have been sensationalized in modern media (and in some cases satirized), their ecstatic and religious aspects have been ignored and suppressed so as to be virtually invisible. If we return to our definition of religious ecstasy for this book, "intense altered emotion, perception, and identity in relation to a religious object," it is clear that there are many forms of ecstasy associated with religious goals in the traditional countercultural "trinity of sin." We can see intense joy and love in music from rave groups to Hip-Hop Hasidim, altered perception in Technopaganism and Urban Shamanism, and altered identity in Trance Dance and the states of unity of the Dances of Universal Peace.

In these cases, we may note that the "trinity of sin" is not an enemy of religion, it is a competitor for the same spiritual audience. However, it understands spirituality in a different way. Their struggle is an extended argument over definition and interpretation.

But as we can see in our next chapter, some of the mainstream religions have had their own reactions to the denial of ecstatic states in their Theology and ritual, which have influenced the development of the millennial, charismatic, and renewal strains in modern American religion, and thus the ecstatic backlash in the West.

Notes

1. Pew Forum, 2012. See Funk, Cary and Smith, Greg. "'Nones' on the Rise: One-in-Five Adults Have No Religious Affiliation" (PDF). pewforum.org. *The Pew Forum on Religion and Public Life.*
2. James S. Gordon, *The Golden Guru: The Strange Journey of Bhagwan Shree Rajneesh* (Lexington: The Stephen Greene Press, 1987), p. 48.
3. Osho, ibid., p. 153.
4. Ibid., p. 114.
5. Bernard Gunther (Swami Deva Amit Prem), *Neo-Tantra*: Bhagwan Shree Rajneesh on Sex, Love, Prayer and *Transcendence* (San Francisco: Harper and Row), p. 47.
6. Advertising pamphlet, Kahua Hawaiian Institute, n.d.
7. From the 1995 brochure, "SkyDancing Institute USA: Tantra Love and Ecstasy Trainings with Margo Anand."
8. Anand, ibid., p. 23.

9. David and Ellen Ramsdale, *Sexual Energy Ecstasy: A Practical Guide to Lovemaking Secrets of the East and West* (New York: Bantam Books, 1993), p. 221.
10. Ramsdale, ibid., p. 223.
11. For greater detail on these practices, see Ramsdale, ibid., pp. 230–231.
12. Ibid., p. 265.
13. Ibid., p. 195.
14. Ibid., p. 333.
15. Ibid., pp. 366–367.
16. From a lecture at the Plattsburg, New York, Sun Myung Moon Center, 1986.
17. Peter Gardella, *Innocent Ecstasy: How Christianity Gave America an Ethic of Sexual Pleasure* (New York, Oxford: Oxford University Press, 1985), p. 10.
18. Ibid., p. 11.
19. Sylvester Graham, cited in Gardella, ibid., p. 57.
20. Ibid., p. 4.
21. Elizabeth Lowell, *Tell Me No Lies* (Ontario: Mira Books, 1986), pp. 40 and 307. The writer is listed as the author of such books as *Granite Man, Outlaw* and *Warrior.*
22. Barbara de Angelis, *Passion* (New York: Delacorte Press/Bantam Books, 1998), p. 62.
23. Diane Ackerman, *A Natural History of Love* (New York: Random House, 1994), p. 98.
24. Ibid., p. 95.
25. John M. Haffert, *Sex and the Mysteries* (Ave Maria Institute, 1970), pp. 134–136.
26. This is a form of spiritualism based on the teachings of the medium Allen Kardec, the "father of spiritism," who wrote during the nineteenth century. His ideas later came to Brazil; he had an emphasis on "spirit communications" and automatic writing.
27. Polari de Alverga, Alex. *Forest of Visions: Ayahuasca, Amazonian Spirituality, and the Santo Daime Tradition*, trans. Rosana Workman (Rochester: Park Street Press, 1999), p. 1.
28. Ibid., p. xxii.
29. Ibid., p. xxiii.
30. Ibid, pp. 54–56.
31. Ibid, p. 57.
32. Ibid, p. 101.
33. Edward Macrae, "The Ritual and Religious Use of Ayahuasca in Contemporary Brazil," in *DPF XII Policy Manual*, ed. W. Taylor et al. (Washington: Drug Policy Foundation Press, 1999), pp. 47–50.

34. Morgan Brent, "Ayahuasca, Religion and Nature" from the Santo Daime Archives, http://www.santodaime.org/archives/morgan.htr.
35. Ibid., p. 3.
36. Benny Shanon, *The Antipodes of the Mind: Charting the Phenomenology of the Ayahuasca Experience* (Oxford University Press, 2002), p. 130.
37. Ibid., p. 113.
38. Ibid., p. 114.
39. Ibid., p. 143.
40. Ibid., p. 130.
41. Ibid., p. 154.
42. Joost A. M. Meerloo, *The Dance: From Ritual to Rock and Roll- Ballet to Ballroom* (Philadelphia and New York: Chilton Company, 1960), p. 32.
43. Gilbert Rouget, *Music and Trance: A Theory of the Relations Between Music and Possession* (Chicago: University of Chicago Press, 1985), p. 92.
44. Cited in Rouget, ibid., p. 220.
45. Kay Gardner, *Sounding the Inner Landscape: Music as Medicine* (Stonington: Caduceus Publications, 1990), p. 129.
46. Ibid., p. 136.
47. Rouget, ibid., p. 126.
48. Ibid., p. 76.
49. Ibid., p. 75.
50. Ibid., p. 79.
51. Ibid., p. 91.
52. Meerlo, ibid., p. 12.
53. Ibid., p. 23.
54. Ibid., pp. 24–25.
55. Iris J. Stewart, *Sacred Woman, Sacred Dance: Awakening Spirituality through Movement and Ritual* (Rochester: Inner Traditions, 2000), p. 184.
56. Ibid., p. 185.
57. Ibid., p. 187.
58. James Hughes, *Creativity Under the Influence* (New York: Watson-Guptill, 1999), p. 27.
59. Tiffany Scott, *Technopagan Sound Excursions: Integrating Tribal Ecstatic Experience through Dance Music Culture* (California Institute of Integral Studies: PhD dissertation, 1998), p. 21. UMI facsimile.
60. Ibid., p. 1.
61. Cited in Nicholas Saunders et al., *In Search of the Ultimate High: Spiritual Experience Through Psychoactives* (London: Rider, 2000), p. 170.
62. Scott, ibid., p. 11.
63. The term is from D. Rushkoff, cited in Scott, ibid., p. 39.

64. Saunders et al., ibid., pp. 181–182.
65. Simon Reynolds, *Generation Ecstasy: Into the World of Techno and Rave Culture* (Boston: Little, Brown and Company, 1998), pp. 238–239. The term is from Antonio Melechi.
66. Scott, ibid., p. 31.
67. Ibid., p. 55.
68. Reynolds, ibid., p. 8.
69. Ibid., p. 125.
70. Ibid., p. 130.
71. Scott, ibid., p. 22.
72. Ibid., p. 34.
73. Saunders et al., ibid., p. 188.
74. Reynolds, ibid., p. 89.
75. Ibid., p. 81.
76. Jerome Beck and Marsha Rosenbaum, *The Pursuit of Ecstasy: The MDMA Experience* (Albany: State University of New York Press), p. 90.
77. Beck and Rosenbaum, ibid., p. 90.
78. Ibid., p. 91.
79. Ibid., pp. 22–23.
80. Richard Rudgley, *The Encyclopedia of Psychoactive Substances* (New York: Thomas Dunne, 2000), p. 92.
81. Ibid., p. 92.
82. Suzanne Smalley, "The New Age of Rave," *Newsweek*, July 7, 2003, pp. 52–53.
83. This information is from a paper by Gordon Lynch, "Contemporary Ecstasies: Exploring British Club Culture as Mysticism" given at the 2003 meeting of the American Academy of Religion.
84. Michael Harner, *The Way of the Shaman: A Guide to Power and Healing* (New York: Bantam Books, 1982), p. xvi.
85. Ibid., p. 39.
86. Ibid., p. 78.
87. Frank Natale, *Trance Dance: The Dance of Life* (Shaftesbury: Element, 1995), p. x.
88. Ibid., p. 50.
89. Ibid., p. 86.
90. Amoda, *Moving into Ecstasy: An Urban Mystic's Guide to Movement, Music and Meditation* (London: Thorsons, 2001), p. 165.
91. Ibid., pp. 181–182.
92. Ibid., p. 182.
93. For the debate, see http://www.spiritvoyage.com/blog/index.php/shiva-rea-and-gurmukh-find-controversy-at-international-yoga-festival-rishikesh/. Also, https://www.opposingviews.com/i/shiva-rea-stirs-up-controversy-at-world%e2%80%99s-biggest-yoga-festival.

94. Michael Drake, *The Shamanic Drum: A Guide to Sacred Drumming* (Bend: Talking Drum Publishing, 1991), p. 19.
95. Ibid., p. 47.
96. Ibid., p. 69.
97. Ash, ibid., p. 24.
98. Gabrielle Roth, *Maps to Ecstasy: Teachings of an Urban Shaman* (San Rafael: New World Library, 1989), p. 1.
99. Ibid., p. 2.
100. Ibid., p. 175.
101. Ibid., p. 61.
102. Ibid., p. 175.
103. Ibid., pp. 201–203.
104. Amoda, ibid., pp. 210–211.
105. Music review of the album "Birds of Fire" by John H. Patterson, December 1, 2000, at http://www.allaboutjazz.com/reviews/r1200_027.htm.
106. Cited in Jon Solomon, "Guitar guru John McLaughlin's spiritual path has been a long one," online at http://www.westword.com/2010-12-02/music/john-mclaughlin-boulder-theater/2/.
107. See http://www.rollingstone.com/music/artists/santana/biography.
108. Interview, *Rolling Stone magazine*, March 16, 2000, p. 89.
109. See http://www.shturem.org/index.php?section=news&id=20932 for an interview with the singer.
110. See the description of the group Petra as http://christianmusic.com/petra/petra.htm.
111. See "Did God Give Rock and Roll to You? Studying the Origins of Christian Rock" at http://www.sinisachoice.com/uploads/Did_God_Give_Rock_and_Roll_to_You_Final.pdf.
112. Cited in Christopher Partridge, *The Re-enchantment of the West: Alternative Spiritualities, Sacralization, Popular Culture and Occulture*, vol. 2 (London: T and T Clark, 2005), p. 100.
113. Nelson, ibid., p. 43.

CHAPTER 7

The Return of the Repressed: Millennial, Charismatic, and Renewal Movements

In the modern world, most religions must live without ecstasy. This is especially true for the more conservative and fundamentalist wings of the three great monotheisms: Judaism, Christianity, and Islam. For all three, there have been times of ecstasy in the past, where God spoke through prophets and did miraculous things on earth. There will also be ecstasy in the future, in the Last Days, with the coming of the Messiah or Christ or the Mahdi. But for now, we live in a time where claims of divine vision, call, and prophetic election are generally forbidden, if not heretical.

There are few ecstatic saints today: Thus, we see the Protestant emphasis on all people being sinners, the Jewish lack of a Messiah, and the recent Wahhabi demolition of Muhammad's old house in Saudi Arabia to make sure that he is not understood as a saint, and his house not an appropriate place for pilgrimage. Even Catholicism has moved from ecstatic and visionary saints to activists for the church, beatified for their charity, good deeds, fund-raising and political abilities, or firmness in anti-abortion statements. Their required number of miracles is shrinking, substituted by "heroic virtue."

Despite the rejection of ecstasy and its associated states, the supernatural realms, we do see religions trying to reclaim what has been rejected. While mainstream Christianity in the USA has largely downplayed the importance of ecstatic experience and focused instead on the ethical and rational aspects of religion, we see two non-mainstream Protestant responses that have become quite important over the past century.

J. McDaniel, *Lost Ecstasy*, Interdisciplinary Approaches to the Study of Mysticism, https://doi.org/10.1007/978-3-319-92771-8_7

One is the Millenialist movement, which emphasizes ecstatic states in the past with the prophets and apostles, and ecstatic joy in the future with the Rapture and the Kingdom (when the saved will be in God's presence). It simply lacks ecstasy in the present. In living in between one of Christ's comings and the next, the Christian obligation is obedience to the moral teachings of the Bible and faith in its revelations. The miraculous events of the Bible stopped when the canon was closed, and must wait until the End-Times to occur. While we might have elements of ecstasy in the conversion experience, leading people to believe in Jesus Christ as their personal savior, its importance is that it brings the person to the correct perspective on salvation. Ecstasy itself is secondary and unnecessary; it is having the right faith that saves. Ecstatic experience will come again in the future, with events like the Rapture. Thus, though we live in a non-ecstatic period of time right now, it is only temporary, and a new day is coming just around the corner with Jesus' return.

This will be happening soon—many people believe it will occur in their own lifetimes, and they will see people levitate into the air and have divine visions and dreams. There are televangelists who argue that the Iraq war was necessary because it was the beginning of Armageddon, and the USA must fight in Israel as well and bring Jews there from all over the world so Jesus will appear.[1] The End is coming so soon that, in the words of some Christian television evangelists, "You will not die!"

The other approach is the Charismatic movement. This approach does not accept the rational and ethical status quo, but rather rebels against it. The canon may have been closed, but the Holy Spirit has not gone into hibernation. Instead, it acts to inspire people by continuing revelation. Its major ecstatic aspect is "speaking in tongues" or glossolalia, in which the Spirit comes down to speak through the person's mouth in an acceptable form of possession. However, there are other ecstatic and miraculous gifts of the Holy Spirit which may occur—faith healing, prophecy, inspired teaching, and new revelations which give insight into older ones. At charismatic church services, we might see ecstatic singing, break dancing, enthusiastic bounding up and down, and being "slain in the spirit." Enthusiasm takes a variety of forms of entranced song and dance. From the charismatic approach, ecstasy never left the world. The Holy Spirit has been with the church (for Catholic charismatics, as the body of all saved people) since the days of the Apostles, and it has simply been obscured by things like the rise of science, and problematic Biblical interpretation. But as the world can only exist through God's grace, human beings can only have the "full gospel" by recognizing the supernatural and miraculous presence of the Holy Spirit as it appears in people's lives.

Both of these traditions have had an important effect on twentieth-century culture in the USA, especially in terms of law and politics.

Millennial Christianity and the Rapture

The modern apocalyptic strains of Christianity, found primarily in the Protestant tradition, predict a dangerous and glorious future for mankind. There will be an ending to ordinary time and space and history, and a set of disasters which will separate out the Saved from the Damned. During this set of events, an ecstatic event commonly known as the Rapture will take place, and people will be transformed, rising into the air to meet with Jesus.

The literature about the Rapture tends to deal with two issues: when and in what historical circumstances the event will occur, and what exactly will happen. This belief derives primarily from Revelation 20:1–6, as well as Enoch, Jubilees, Esdras, and Daniel. There are also references used from Paul's letters. There is no single, definitive theologian of the Rapture. It has been mentioned in sermons by evangelists like Oral Roberts and Jack Van Impe, and there have been fiction books about the Tribulation period, of which the most well known is the Left Behind series by Tim LaHaye and Jerry Jenkins. The most quoted writer is probably Hal Lindsay, author of several popular books on apocalyptic ideas of the future, the most famous being his 1970 book *The Late, Great Planet Earth*.

While earlier understandings of the Revelation of John assumed it to symbolize events in the past, more recently it has been understood as predicting the future of the earth. Oral Roberts, in his *The Drama of the End-Time*, spoke of two major events that would happen at the end of the world—the Rapture and the Revelation. He described the Rapture as the "catching away of the bride," with Christ suddenly appearing for a minute in the air to take away the members of his church. The Revelation would include the Battle of Armageddon and the Millennial Reign. Roberts was interested in the flight of people into the air, but his focus for this was upon the resurrection of the dead rather than the Rapture. As he writes,

> And from all over this old sin-cursed, Devil-ridden, hell-bound world, the Bible-reading, blood-washed, spirit-filled, Christ-loving, God-serving, faithful band of holy people, from Abel's day to this, shall bid goodbye to earth's sorrows and woes... What a flight! the rapid release of the underground Minuteman missiles and the underwater Polaris submarines are not to be compared with the instant rising of the dead in Christ from graves under the earth and sea.[2]

There would be both Rapture of the living and resurrection of the dead, and in this second flight, the redeemed would be resurrected and translated, joining the marriage of Christ and his Church which would go on in heaven. However, the literature has shifted from an emphasis of the flight of the dead to an emphasis on the flight of the still living.

Most modern ideas of the Rapture are found within the theological perspective of dispensationalism. This says that God works in history in various dispensations, or historical eras, and in its millenialist aspects says that that Jesus' triumphant return to earth will result in a thousand-year earthly reign. It is part of the theological study of salvation history, the ways that God works in the world. The timing of the future historical events associated with the Rapture is debated in the interpretations known as pre-millenialism, post-millenialism, and amillenialism. The pre-millenialist understanding is that "saved" or faithful Christians will undergo the Rapture in relation to a set of historical disasters known as the Tribulation. The Tribulation will last for a set period (usually seven years), until Jesus returns to the earth to set up a thousand-year kingdom. Pre-Tribulation Rapture has people saved before these disasters, mid-Tribulation Rapture has them saved in the middle of this period, and post-Tribulation Rapture has the saved Christians suffering through the Tribulation, but then being "raptured up" at the time of Jesus' return to earth.

More specifically, the pre-Tribulation Rapture involves seven years of misery before Christ returns, with the church raptured or resurrected before this point. They will be taken and judged, and then celebrate the marriage of Christ and the Church, and the marriage supper of the Lamb. After seven years, Christ will come to earth, destroy the Antichrist, chain Satan in a pit, and set up his kingdom. In the mid-Tribulation view, the church experiences three and a half years of human wrath, but avoids "the wrath of God." In the post-Tribulation view, the church endures the Tribulation together with all of humanity, and the

Rapture occurs with the other end-time events, like Christ's appearance and judgment. There is also a "partial Rapture" approach, which distinguishes between classes of Christians. From this perspective, the most spiritual Christians will be taken before the Tribulation events, while less spiritual ones will be taken after the Tribulation (this approach thus combines pre- and post-Tribulation approaches). All of these are approaches of the pre-millenialist viewpoint, which is currently the dominant one in apocalyptic (or prophetic, as it is sometimes called) Christianity.

The minority approaches are post-millenialism and amillenialism. They understand both the Rapture and the Tribulation to be symbolic rather than literal events. The former holds that the gospel and the church will be triumphant, and a millenium of peace or golden age will come. Jesus will return at the end of this period. The latter holds that there is no millenium period, but that Jesus will come at the end of the world.[3]

The interpretation of the Rapture in pre-millennial eschatology has a debated past. There is general agreement that it was described in the early nineteenth century by John Nelson Darby, an Anglican priest and the leader of a group called the Plymouth Brethren. Darby spoke of two comings of Christ—in the "secret rapture" of the church that would occur first, and later in the public Second Coming of Christ. This approach, at the time called "futurism," later came to be called "dispensationalism." It was then popularized in the USA in the later nineteenth century by C. I. Scofield, who wrote the Scofield Reference Bible, and later followed by Dwight L. Moody of the Moody Bible Institute, and by evangelists like Billy Sunday.

It has been suggested by J. Preston Eby (and stated on a number of current Protestant Web sites) that the origin of the Rapture itself was in the writings of a Jesuit priest named Ribera, as part of the Catholic response to the Protestant Reformation. Ribera wrote that all events in the Book of Revelation would take place literally at the end of the age. He stated that these ideas were elaborated by another Jesuit named Emmanuel Lacunza, writing under the name Rabbi ben Ezra (ostensibly a Christian convert). Lacunza wrote that Jesus would return twice, the first time to rapture away Christians, and the second time to deal with the Antichrist (thus there is both a second and a third coming). This Catholic origin of the Rapture has been problematic for some Protestants.[4]

In the nineteenth century, the preacher Edward Irving read Lacunza and preached the Secret Rapture of the Saints. He translated Lacunza's work into English, and his preaching inspired the Plymouth Brethren. The ideas of Irving and Darby were also affected by the visions of a teen-aged girl, Margaret Macdonald, in 1830 (though Hal Lindsay disagrees with her importance, saying that she only described a partial Rapture, and not a full one).[5] Its most controversial claim is that Jesus does not return once, for the Second Coming, but twice, and that the public Second Coming is really a Third Coming.

What was understood to happen at the Rapture? For those living during that period, it means that they will not die and will be taken up and transformed into immortal beings. This is understood to surpass the experience of the Biblical figure of Elijah, who was taken off to heaven in a fiery chariot. Sometimes Jesus is also considered to be a Biblical precedent for the Rapture, as he died but then continued to live. Immortality is an important issue here. As Lindsay states:

> … the meaning of the mystery is that we Christians are not all going to die! This was a totally new concept. No Old Testament believer dreamed that some future generation would enter eternity and God's presence without experiencing physical death. Death is an absolute prerequisite to entering immortality through resurrection… The truly electrifying fact is that many of you who are reading this will experience this mystery. You will never know what it is to die physically.[6]

This change or transformation into immortality means conformity with Jesus' body of glory or resurrection body and will occur in the twinkling of an eye, as Paul phrases it, at the last trumpet (1 Cor 15:52–53):

> Listen! I will unfold a mystery: we shall not all die, but we shall all be changed in a flash, in the twinkling of an eye, at the last trumpet-call. For the trumpet will sound, and the dead will rise immortal, and we shall be changed. This perishable being must be clothed with the imperishable, and what is mortal must be clothed with immortality.

The expected process is described in greater detail in Paul's First Letter to the Thessalonians (1 Thes 4:15–18):

> For this we tell you as the Lord's word: we who are left alive until the Lord comes shall not forestall those who have died; because at the word of

> command, at the sound of the archangel's voice and God's trumpet-call, the Lord himself will descend from heaven; first the Christian dead will rise, then we who are left alive shall join them, caught up in clouds to meet the Lord in the air. Thus we shall always be with the Lord.

The term for "caught up" is the Greek verb *harpazo*, literally meaning "to snatch" or "to seize." It later becomes "rapture," taken from Jerome's Latin translation of the term, meaning to be "caught up." The Christian is grabbed by God, and all saved persons will undergo an instantaneous transformation into immortals, gaining a body like Jesus' glorified body. This will occur in the air, and there is some debate as to whether this levitation will be temporary or long term (for the years of the Tribulation). As Lindsay notes, "Just think of it… in the flash of a second every living believer on earth will be gone. Suddenly, without warning, only unbelievers will be populating planet earth."[7]

Some writers interpret being seized by God to mean that the transformed or translated souls will then go to live in a spiritual abode in heaven, until a mystical union with Christ. Others say that immediately after meeting God in the air, the levitating people will return to earth with Christ. All agree that believers will gain immortal bodies.[8]

The terms which are traditionally used to refer to the Second Coming of Christ are variously interpreted to refer to the Rapture by different modern authors; these include such Greek terms as *parousia, apokalypsis*, and *epiphaneia*. Sometimes these are understood as a separate phenomenon from the Second Coming, and sometimes as an early stage of the event.

The Rapture is the "crown of exaltation," the greatest joy a person can know. It is gaining perfection and glorification, and later a special kind of union. The Secret Rapture also has the romance of the unknown and mysterious, the consciousness of being saved and in a special class which is different from the average sinner, and gaining a great reward for difficult religious behavior. The popular literature on the Rapture includes freeways full of cars whose drivers have suddenly disappeared, manhunts, graves whose bodies have disappeared, babies disappearing from their mothers' arms. While popular books about the Rapture include much drama, they are often surprisingly lacking in charity and compassion for non-Christians.[9]

Such ideas as the Rapture and Tribulation are controversial within modern Christianity, and even within various strains of fundamentalist Christianity. In books like Tim LaHaye's *The Rapture (Under Attack)*, there is much infighting between prophetic groups about when the Rapture will come (despite Jesus' rather clear statement that nobody knows when it will come, in Mark 13:32, and Matthew 25:13). LaHaye notes that the importance of belief in his interpretation of the Rapture is that it produces a more evangelistic church and more missionary work,[10] a pragmatic argument for supernatural belief. He notes his credentials: "I myself have been a forty-five year student of the satanically-inspired, centuries old conspiracy to use government, education and media to destroy every vestige of Christianity within our society."[11] He believes this conspiracy to be based on plots by the Illuminati.

Here, we characterize the Rapture as a form of ecstatic experience because it is understood to happen to living human beings, and is not an after-death event. It includes altered emotion (intense joy at being with Christ), altered perception (looking down from up in the air), and altered identity (glorification and perfection, becoming like Christ, and gaining immortality).

It is also a form of resurrection which occurs to living people. Millennial thought includes two categories of resurrection, the "resurrection of life" and the "resurrection of damnation" (from John 5:29, when "those who have done right will rise to life; [and] those who have done wrong will rise to hear their doom"). The first resurrection is understood to refer to the Rapture, and the second resurrection refers to the judgment of unbelievers before the great white throne at the end of time (Rev 20:11-15), when Death gives up its dead, and people undergo a judgment and in some cases a second death in a lake of fire.

There is some debate about whether the Rapture is really secret and silent, as the Revelation of John describes the end-times as full of trumpets and shouting. There is debate about who is saved when, and what follows. But it is clear that this is an ecstatic state greatly to be desired, and there are many millenialists today who believe that they will not die, but be transformed into joyous and levitating people during their lifetimes. Ecstasy is the mark of a glorious future.

Charismatic Christianity and the Baptism of the Holy Spirit

Charismatic Christianity is one of the fastest-growing forms of Christianity in the twentieth century, becoming very popular in Latin America, Africa, and Asia. Its most important ecstatic experience is the

Baptism of the Holy Spirit or Fire Baptism, which is understood to follow after ordinary baptism. In this event, the person is possessed or taken over by the third member of the Trinity, who brings intense joyful emotion, altered perceptions, and a sense that the body and mind are being controlled by some greater force. In his book *The Inner Ecstasy*, Marcus Bach gives a detailed description of his own experience:

> My body was shaking. I was hot and cold. An ecstatically pleasant, thrilling surge of passion swept through me as if the Holy Ghost, whatever it was, had finally found an entry and was rushing in to take control. The shaft of light was real and visible behind my tightly closed eyes... It was insanely wonderful. For the first time in my life a liberty beyond understanding and a joy that knew no bounds had hold of me. Just to lie on the floor and roll on the floor and shout prayers to God in a voice and words that only he could recognize, that only He could have put in my heart in the first place...the outpouring of heavenish words, the volleys of Spirit-loaded phrases washed out every frustration and barrier that had ever kept me from saying, "Lord, I believe" I was saying it now in a "language of the Spirit" of which I understood not a single word. Mine was a feeling, a feeling that I was being *spoken through*... My words were drowned in a flood of ecstatic laughter.[12]

Baptism of the Holy Spirit is usually associated with speaking in tongues, as this quotation describes. However, sometimes it comes later in a delayed reaction, as Bach cites from a letter from a friend:

> I received the Holy Spirit from the laying on of hands. I collapsed on the floor and it felt like about 500 volts. There was a spasm of pressure and violet and red light. I nearly lost consciousness. For a week the world seemed different and there was an amazing alignment with all things, a synchronicity... I did not speak in tongues, but several weeks later I was talking to a minister over the phone and he said, 'You can speak in a tongue right now!' I laughed him off and hung up. I was alone. I stood up and decided to try it anyway... It was a powerful and melodious language.[13]

A whole set of ecstatic events is understood to accompany the baptism, such as the ability to speak in tongues, to prophesy, to discern spirits (determining which are good and which are evil, along with gaining the ability to exorcise the evil ones), to gain supernatural wisdom and knowledge, the ability to heal by prayer, and others. While more formally these originate in Paul's First Letter to the Corinthians in the New Testament, other modern gifts of the Holy Spirit seem to follow more informally, such as holy laughter, ecstatic dance, and being slain in the spirit.

Baptism of the Holy Spirit is described in several places in the New Testament, most notably Acts II:1-4:

> While the day of Pentecost was running its course they were all together in one place, when suddenly there came from the sky a noise like that of a strong driving wind, which filled the whole house where they were sitting. And there appeared to them tongues like flames of fire, dispersed among them and resting on each one. And they were all filled with the Holy Spirit and began to talk in other tongues, as the Spirit gave them utterance.

As this event took place at Pentecost, people who emphasized its importance later came to be called Pentecostals. Because it was accompanied by such *charismata* or spiritual gifts as speaking in tongues, we also see the term "Charismatic" used. In recent years, the term "Charismatic" has become a more popular term for the group than Pentecostal.

Speaking in tongues or glossolalia is a well-known ecstatic religious phenomenon—it was present at the ancient Greek Eleusinian and Dionysic festivals, and seen in the seeress or Pythia of Apollo at Delphi (who had priests present to translate), and among the groups of ecstatic prophets described in the Hebrew Bible. It is often found in both individual and group trance states and is relevant to the figures valued by those groups—Mormons spoke Indian dialects, and Spiritualists spoke in the languages of ghosts and spirit guides. The early experiences of George Fox and his followers included speaking in tongues along with the visions, groaning, prophecy, and the quaking or trembling that gave the Society of Friends their nickname of "Quakers." However, glossolalia became especially important among Charismatics, as the most important mark of the descent of the Holy Spirit.

Baptism of the Holy Spirit is described as intense, joyful, ecstatic, overwhelming. Pentecostal and Charismatic Christians describe it as "full consummation" when the Holy Spirit takes full control of the believer's body, soul, and spirit. While they follow the traditional Pauline tripartite description of the person as body, soul, and spirit, there is often suspicion of theological anthropology among Charismatic Christians, and the

Baptism is described in rather vague terms, theologically speaking. One analogy comes from the Assemblies of God:

> As Christians we may be compared with a reservoir for producing electrical power. When we accept Christ, construction of our reservoir is complete. We now have the potential to be useful and to affect lives. But until the flood gates are opened and the cascading river waters pour through, no power is realized. So it is when we are baptized in the Holy Spirit. We open our lives to God and the Holy Spirit pours into us and through us... when our spiritual power runs low, we need to return to the Source and let the blessed Holy Spirit pour into us again, bringing fresh power.[14]

Sometimes the person is a temple, sometimes a reservoir, and the spirit is regenerated while the soul and body are immersed, dipped, sunk, or plunged. There are many liquid metaphors used. Usually, the body is filled by the Holy Spirit, which "wells up" but "comes upon" the soul. The body may be described as an empty vessel, a jar, a vase, or a bowl filled by the Holy Spirit. There is overlapping use of the terms soul and spirit—usually the soul refers to life, mind, or personality, while the spirit is associated with breath, the inner person, or the intellect, and sometimes the image of God within the person. This uncertainty may reflect the Bible's ambiguous uses of the words "soul" and "spirit," for it does use varying referents for these terms.

Charismatic Christianity is largely a twentieth-century phenomenon, expanding from the Pentecostal movement into other forms of Christianity (today there are Charismatic movements in Catholicism and in many Protestant denominations). There are also subgroups, such as the Vineyard, Word-Faith, and Holy Laughter movements.

We do see earlier examples of charisms in Christianity, which were mentioned by such Church Fathers as Irenaeus, Tertullian, Augustine, and St. John Chrysostom. However, the major precursor to modern Charismatic Christianity is probably early Methodism and its revivals.

The early Methodists had a strong ecstatic tradition, feeling that holiness or entire sanctification was possible for people, if they submitted themselves to the Holy Spirit. There was the stage of conversion (salvation), followed by complete holiness (sanctification), also called the Holy Ghost Baptism or Spirit Baptism. During the Great Awakening in Virginia, Jarratt notes in his 1775 *Brief Narrative of the Revival of Religion* that the holiness tradition brought all sorts of emotional responses:

> Some would be seized with a trembling, and in a few moments drop on the floor as if they were dead; while others were embracing each other with streaming eyes, and all were lost in wonder, love and praise.[15]

Methodist sanctification was often an intense experience, and this approach grew into the "second blessing" tradition and the Holiness Movement. The seventeenth through nineteenth centuries had varied forms of religious enthusiasm in England and colonial America. Revivals spread in the eighteenth and nineteenth centuries on the American frontier, emphasizing miracles, healing, exorcism, and speaking in tongues, and preachers often demanded a tangible display of the receiving of the Holy Spirit.

The "second blessing" came to have a number of synonyms: the Baptism of Burning Love, the Baptism of the Holy Ghost and Fire, and the Fire Baptism. By the nineteenth century, those receiving "the Fire" in the Holiness tradition would experience a variety of states: they would have the holy dance and holy laugh, speak in tongues, shout and scream, and get "the jerks." As G. F. Taylor describes, these events in 1895:

> Some said they felt the fire burning in their souls, but others claimed it as burning in their bodies also. It was felt in the tongue, in the fingers, in the palm of the hand, in the feet, in the side, in the arms, and so on. The Bible itself often felt warm to those who had the fire in them. The church would seem to be lighted with fire, the trees of the wood would appear as flames of fire, the landscape would seem to be baptized in the glory of the fire. As one rode from one appointment to another, according to their testimonies, they seemed to be enveloped with the holy fire. The noise of the engine seemed to sound notes of praise to God, and the clatter of the wheels beneath the cars seemed to be saying, *Glory to God, hallelujah!* The coaches themselves were fire-lighted, and the wheels beneath seemed to be wheels of fire. Fire! Fire! Fire! was the ring of their testimonies.[16]

The Kentucky Revival of the nineteenth century also had a range of charismatic experiences, especially the "falling exercise" and those who were "stricken of the Lord" and lay on the ground as if they were dead (these later came to be known as being "slain in the Spirit"). There were hundreds of people who lay on the ground for hours. There were also people who had the "barks," and barked like dogs or imitated other animals, and the "jerks," in which case their bodies were convulsed and moved involuntarily, twitching, hopping, and jolting. The barks were involved in "treeing the devil," in which people would crawl on all fours and bark. There were also the "rolling exercises" in which people doubled over and rolled around like wheels or stretched out and turned over like logs, and "such wild dances as David performed before the Ark of the Lord."[17] There was holy laughing, shouting, and song.

Such phenomena were not limited to the Holiness traditions—Shakers had the "whirling gift" in which they spun like tops, the "dumb devils" in which they were unable to speak, the "mortification gifts" which included making animal sounds, and the "laughing gift" in which their laughter was understood as a holy blessing.

Many of these experiences have come to be experienced in the twentieth-century Charismatic movement and understood as a new set of anointings or blessings. However, the emphasis came to be on speaking in tongues rather than the other gifts of the Spirit. We see this especially in the early Pentecostal movement, which E. Glenn Hinson describes as partial reaction to the increase of secularism and the waning of revivalism following the Civil War.[18]

The Pentecostal or Charismatic movement is usually considered to have begun in 1901 with Charles Fox Parham, a Methodist preacher in Kansas. He adopted the teachings from Benjamin Irwin's Fire Baptized Holiness Church, which included three important experiences: Salvation (by water baptism), Sanctification, and Baptism of the Spirit or Baptism of Fire. Parham spoke in tongues, as did the people at his Bethel Bible school, and he taught that this "Second Pentecost" or Baptism by the Holy Spirit was necessary, and speaking in foreign tongues was its evidence. He predicted that there would be a worldwide Latter Rain revival, when the Holy Spirit would bring back the gifts of the Spirit to create a great End-Times Harvest, at the end of which Christ would return to earth.

When one of Parham's students, a black Holiness preacher named William Seymour, brought Parham's ideas to Los Angeles in 1906, he started an ecstatic movement which came to be called the Azusa Street Revival. It included the "jerks," the "holy laugh," the "holy dance," and singing in the spirit.[19] Crowds wept, cried out, and felt great religious excitement. They were slain in the spirit "like warriors in battle," cried for hours on end, spoke, and sung in tongues (creating a "heavenly chorus") and underwent healing though atonement for sin and Jesus' blood. Mixed racial and ethnic groups shouted and cried in expectation of the

Second Coming. As these ideas spread, there were reports of raising the dead, preachers levitating, balls of supernatural fire, bloody crosses appearing on peoples' foreheads, and "Holy Ghost Oil" appearing on their hands.[20] The ecstatic dancing, fainting, and rolling on the ground gave them the nickname of "Holy Rollers."

Over the following decades, many "full gospel" and "full consummation" churches arose, and there were charismatic preachers and independent evangelists: Oral Roberts, Larry Lea, Kenneth Copeland, Bob Tilton. In about 1960, the "neo-Pentecostal" movement began when two Episcopalian priests began to speak in tongues, and in 1967, this phenomena was accepted by the Catholic Church.

Also in the 1960s was the rise of the Jesus People on college campuses and high schools, and there were students performing glossolalia on campus (called "glossoyalies" at Yale). Some of these people had formerly been a part of the drug culture, and converted after encountering Christian communes, Gospel rock, and coffeehouse ministries. Others were more generally a part of the Born Again movement, which placed emphasis on people having a personal relationship with Jesus (and which might or might not involve other charismatic gifts).

Currently, there are a variety of sub-movements within the Charismatic Churches. Some hold to the doctrine of "restorationism" or "Latter Rain restorationism," where God's power will enter the churches to strengthen them before the Second Coming, and the Holy Spirit will affect the world through signs and miracles, and especially through new orders of apostles and prophets (though the Latter Rain philosophy was rejected in 1949 by the Assemblies of God General Council).[21] Another sub-group is the Word-Faith/Positive Confession teaching, which combines divine healing, spiritual gifts, faith, and financial prosperity. There are also the Manifested Sons of God, who reject Protestant denominations as pagan and believe that the saved will all become "Christs in the flesh," manifesting Jesus on earth, and the Shepherding movement, which emphasized discipleship and absolute submission to church elders in an authoritarian chain of command.[22]

A recent branch within the Charismatic circles is the Vineyard movement, which adopted the "laughing revival" in 1994. A Vineyard pastor preaching in Toronto, Canada, had a congregation in which unusual events happened. There was extended laughing ("holy laughter"), spiritual healing, people being "slain in the Spirit" and imitating animals. People would lie on the floor convulsed with laughter for hours.

They would roar like lions or try to fly around the room like eagles, perform unknown dances or be stuck to the floor with "Holy Ghost Glue." People barked like dogs and crowed like chickens.[23] This became known as the "Toronto Blessing," and it has become very popular, with thousands of people attending these services. However, there has also been disagreement over people acting like animals, and the Toronto Church has separated from the Vineland churches over the "zoo anointing."

Another ecstatic event, the Pensacola Outpouring of 1995, involved people in the audience trembling, collapsing, and remaining unconscious for hours. Participants reported visions of angels dancing in the auditorium and blue hazes, people were imitating animals, and there was twitching, memory loss, feelings of invisible currents of energy, and women simulating childbirth (this was called "fetal birthing" or "birthing a revival").[24] In 1997, a new approach was added at a Chesterfield meeting by the evangelist Kenneth Hagin, the descent of the "Serpent Spirit," where he and his congregation hissed and slithered like snakes.

As we can see, the range of ecstatic phenomena included as gifts of the Holy Spirit has expanded since Paul's time. As Paul states in 1 Corinthians 12:4–11:

> There are varieties of gifts, but the same Spirit. There are varieties of service, but the same Lord. There are many forms of work, but all of them, in all men, are the work of the same God. In each of us the Spirit is manifested in one particular way, for some useful purpose. One man, through the Spirit, has the gift of wise speech, while another, by the power of the same Spirit, can put the deepest knowledge into words. Another, by the same Spirit, is granted faith; another, by the one spirit, gifts of healing, and another miraculous powers; another had the gift of prophecy and another ability to distinguish true spirits from false; yet another has the gift of ecstatic utterance of different kinds, and another the ability to interpret it. But all these gifts are the work of one and the same Spirit, distributing them separately to each individual at will.

We would have to note that many modern additions are not present in Paul's writing, or perhaps might go under the general "miraculous powers." There is no requirement here for speaking in tongues or "ecstatic utterance." Later in First Corinthians, Paul specifically says that ecstatic speech is good for the individual, but not that it is necessary—indeed, he says that love and prophecy are much more useful, as other people can understand them (I Cor 14:1–6).

However, he does value the descent of the Holy Spirit, and he distinguishes between people being baptized and receiving the Holy Spirit (for instance, in Acts 8:5–17, the Samaritans are baptized but receive the Holy Spirit later, and in Acts 19:1–9, the converts at Ephesus had been baptized, but when Paul laid his hands on them, they felt the Holy Spirit and "spoke in tongues of ecstasy and prophesied.") These events happened after the Pentecost of Acts 2 was over.

There are many debates today about the legitimacy and value of charismatic phenomena. These include whether such spiritual gifts should continue past New Testament times or whether they were limited to the primitive church (as a part of the debate over continuing revelation); whether spiritual gifts should only include the gifts mentioned in First Corinthians, or whether new gifts come with new times; and whether churches should follow traditional leadership or "spiritual anointing" (in other words, whether people who demonstrate ecstatic phenomena should therefore take positions of power in the church). People also debate whether one should passively wait for the Holy Spirit to descend ("tarrying upon the Lord") or invoke it in a more active manner, calling God down by laying on of hands. Is baptism sufficient for salvation, or is some more advanced supernatural empowerment of second blessing or "special assurance" of salvation needed?

Indeed, spiritual gifts do not always work as expected. When Charles Parham in 1901 experienced speaking in tongues, he assumed that the Holy Spirit would give his followers instantaneous knowledge of foreign languages if they went out to other countries to preach. He based this on the description of Pentecost in Acts 2, in which people from other countries could understand the disciples as speaking foreign languages when they spoke in tongues. His followers did go out to be missionaries to Japan, China, and India, but despite their faith, they were unable to spontaneously speak their languages of these countries.[25]

For over fifty years, sociologists have been investigating glossolalia in Charismatic Churches, taping speakers to see if they really have been speaking known languages. No instance has been reported of a recognized language being spoken (despite a strong Christian folk tradition which asserts this). As Hinson notes, there are occasional words or brief phrases from other languages, but no evidence that a person had actually developed the ability to speak another language.[26] Today, tongues are popularly described as a private prayer language, or a "language of the angels" which may convey divine messages to a congregation, if an interpreter of the tongues is present.

As noted earlier, charismatic experience has become an accepted event in some Catholic Churches. It is described as echoing an early rite of Christian initiation, involving both the "bath of regeneration" (baptism) and the reception of the Holy Spirit, associated with the proclamation of the death and resurrection of Jesus.[27] Sometimes the emotional nature of the Baptism is emphasized, as Father Edward O'Connor notes in *Ave Maria* magazine:

> The peace is deeper than any they have ever known; the joy wells up from within like a fountain that is ever fresh, and yet it is without the exuberance or elation that makes human joy liable to excess and then deflation.[28]

However, other Protestant writers see it primarily as a response to God's presence:

> "Tongues" is not an emotional evocation of an experience. Rather its nature is response. It is the child's delight, the glee that greets the fireworks display on the Fourth of July, not the display itself.[29]

This experience of the Holy Spirit thus unites Catholicism with a variety of Protestant traditions, including the Quakers and Shakers, Methodism and the Holiness movements, and the modern forms of Pentecostalism.

Can the Baptism of the Holy Spirit be called religious ecstasy? There is intense emotion described, as joy, bliss, and delight, resulting from the presence of the Holy Spirit. There is altered perception, as people have visions of light, fire, colors, the flowing of love, and angels. There is altered identity, as people are possessed by the Holy Spirit, which takes over their physical actions (especially speaking), resulting in glossolalia, being slain in the spirit, and identifying with birds and animals.

For the most part, charismatic religion has been controversial in both the mainstream denominations of Protestant Christianity and also in its more fundamentalist wings. In spite of Christ's statement in Matthew 6 that prayer should be done silently and outside of the social arena, we see a very public display of prayer and ecstasy in many charismatic groups. This is often associated with the coming of the Last Days.

Some forms of modern Protestant Christianity are fundamentalist, believing in a literal and historical understanding of the Bible, and are also apocalyptic, assuming that the Last Days are coming soon, and that people must prepare for them. Charismatic events have been predicted for the Last Days—a portion of the Spirit will be poured out, and young men will see visions, and old men will dream dreams—and these will predict or accompany the sufferings of a future period of Tribulation.

While both millennialism and charismatic beliefs have been present in Christianity since the early church, they have been less visible with the rise of philosophical Theology and the focus on a stable and consistent set of Christian churches, which emphasize ethics and sober behavior. We have a God of Order, who has chased disorder into the shadows. But both survivalist fears for the Last Days and charismatic epidemics of laughing, crying, and prophecy are nothing if not disorderly. These are two forms of religious backlash, bringing chaos and change, yet also spontaneity and hope for ecstatic vision into the churches.

Other Forms of Western Religious Backlash

There has long been a tradition of non-Christian minority religions in the USA, from Eastern religions to groups like Spiritualists and Theosophists, which emphasize religious experience. However, there have also been recent religious responses to the ecstatic drought in the USA from the other traditional Western religions, Judaism and Islam. In American Judaism, we have the Jewish Renewal movement, which has sought both to revive ecstatic rituals from the Jewish past, like prophecy and Kabbalistic practice, and to adopt and integrate practices from other religious traditions (including Zen Buddhism and Native American religions).

We also see the presence of ecstatic Islam in American Sufi groups, especially the Sufi Order of the West. It arose out of the South Asian Chishti order, a Sunni Sufi order which began in the twelfth century. Its modern form was developed by Hazrat Inayat Khan and continued by his son Pir Vilayat Khan. This is a universalist form of Sufism, which accepts the legitimacy of other religions, and has developed a universal worship ritual for devotion to other gods and traditions. Sufism has become popular through Sufi poetry, especially that of the Persian poet Rumi. While it is the more aggressive fundamentalist and mainstream types of Sunni Islam that are more well known, there are Sufi groups in the USA with a great interest in ecstatic states and love of God.

In modern Judaism, the Jewish Renewal Movement is a loose confederation, made up of a diverse group of synagogues, yeshivahs, rabbis, chavurot, networks, minyans, and congregations. It promotes Jewish spiritual and religious experience by embracing the Jewish mystical tradition of Hasidism, and adapting it for the modern world. There is no centralized authority, though many of the leaders have gone through rabbinical programs in a variety of forms of Judaism (primarily orthodox, conservative, reform, and reconstructionist) and had personal training with Hasids, Sufis, Zen masters, Yoga teachers, and other religious practitioners. The general goal of Jewish Renewal is to return Judaism to its experiential roots, and to bring unity and healing to the world (*tikkun olam*). It has been influenced by Jewish mysticism and by various rabbis and practitioners of Judaism who have sought to expand the limits of the traditional religion. While Rabbi Aryeh Kaplan has emphasized the role of meditation and personal practice, creating awe and mindfulness while performing mitzvoth and having an intense experience of solitariness to reach the inner prophetic voice, Rabbi David Lerner has focused upon the roles of spiritual politics, ecology, and social action as ways to help the world.

Jewish renewal is a major contemporary spiritual and ecstatic movement and is an outgrowth of the Havurah or "commune" movement that began in the 1960s. As David Ariel states, it "combined elements of Hasidic ecstasy and communal intimacy in a non-institutional and egalitarian culture."[30] The Havurahs influenced Rabbi Zalman Schachter-Shalomi and Rabbi Shlomo Carlebach, who are often considered founders of the Renewal movement. Rabbi Schachter-Shalomi emphasized the importance of Hasidism for modern Judaism and incorporated practices from other ecstatic traditions to create syncretistic practices. Rabbi Carlebach added Hasidic music and liturgy (as Ariel states, these are "Jewish melodies that induce ecstasy and self-transcendence").[31] Both ecstasy and community are important, as Ariel further states:

> To Jewish Renewal adherents, Judaism without ecstasy is empty, and the support of a community of people connected together by a common spiritual outlook is indispensable.[32]

One important renewal organization, Aleph, was founded in 1976 by Rabbi Zalman Schachter-Shalomi. He was born in Poland, educated in an Orthodox yeshivah, and joined the Chabad Lubavitch Hasidic group. He left the Hasidic community after experimenting with LSD and developed a form of Judaism with an emphasis on meditation and contemplation. He founded a group known as *B'nai Or*, meaning the "Children of Light," a title he took from the Dead Sea Scrolls, and he was known to his followers as the "B'nai Or Rebbe" (with the rainbow prayer shawl he designed for his group called the "B'nai Or tallit"). He added such innovations as "psycho-halakha" (or "integral halakha," a mystical interpretation of the Jewish law), eco-kashrut (kosher laws the follow ecological concerns), contemplative davening or bowing during prayer, and spiritual eldering which emphasized the importance of meditation in old age.

He founded the ALEPH Ordination Programs and ALEPH: Alliance for Jewish Renewal. The seminary he founded has ordained over eighty rabbis and cantors, and it has about forty affiliated communities around the world. Aleph defines Jewish Renewal in this way:

> Jewish Renewal is a worldwide, transdenominational movement grounded in Judaism's prophetic and mystical traditions.. [it] seeks to bring creativity, relevance, joy and an all-embracing awareness to spiritual practice..."[33]

People can learn about God experientially in their "Sage-ing Program," or at C-DEEP, the Center for Devotional Energy and Ecstatic Practice. Members perform meditative practices incorporated from a variety of traditions.

In the Jewish Renewal movement, there is a re-envisioning of the religion, as emotional, populist, existential, even anti-scholastic. There are varying theologies accepted. One widespread approach states that God is within the person, especially accessible as the feminine divine presence or Shekinah. God may be understood in either gender and is not limited to traditional Jewish Biblical and Talmudic interpretation. God continually renews creation and is present within the earth. We see God worshipped in feminine form, and by women who are interested in both prophetic and meditative approaches, for there is an outspoken emphasis on contemporary feminist spirituality, and the current age is one when "the feminine is emerging."[34]

One example of a Renewal school of Jewish prophecy and meditation is the Nefesh Haya School of Ophanim. The Ophanim practices involve yogic meditations, whose origins are attributed to Abraham. A form of hatha yoga is performed in which practitioners stand or dance in positions based on the letters of the Hebrew alphabet. This may be done in individual solitary practice, or in groups, occasionally leading to "Torah raves" in which people take on these positions spontaneously.

The goal is *tikkun ha nefesh*, uniting body, soul, and spirit, until the body is the direct expression of the soul and can speak the word of God through prophecy.[35] While the transmission of this practice in Jewish history has been a bit unclear, the practices are believed to have been originated by Abraham and continued and elaborated by the Baal Shem Tov (founder of the Jewish Hasidic movement), Rabbi Nachman, the Breslover Hasidim, Shlomo Carlebach, and Rav Gedalyah-Koenig. This mythic origin was revealed to Shoshana Weingrove, acting as a modern prophetess.

According to Weingrove, at the beginning of the universe we have the "breaking of the vessels" (a creation story described in Lurianic Kaballah). The spiritual elements and qualities created by God fell from the heavenly realms when the vessels or Sephiroth exploded, and came down to the earthly realm, where their new containers were Hebrew letters. Thus, divine energy fell into the world and entered language, and the Kaballistic goal has been to bring the divine light back to its origin by meditation upon the Hebrew language and letters, and understanding their true meaning. This fall is reflected in the Biblical story of the Tower of Babel, where the division of the original language into different tongues represents the secularization of language. While previous Jewish teaching emphasized restraint, and purification of the physical world by acting only as commanded by God, this new teaching says that Jews are not condemned to spirituality via "not-doing." The body can speak the words of God now and act as a direct expression of *neshamah*, the aspect of the soul united with God.[36]

Before the energies of the Sephiroth were trapped into the Hebrew alphabet, they had appeared to mankind as bodily forms, postures which could dance out messages from God. Once traditional Judaism determined how they might acceptably be approached, people were then limited to finding out about God through letter and text. Now that there is a new revelation, the previous form of Jewish exile (mental exile, or exile in the head away from the body) is over, and Jews can return to find God through their bodies and in the world.[37] The Ophanim practice also involves visualization of the Kaballistic angels in parts of the human body, much as the Hindu Shakta tantric practice of *nyasa* involves the placement of tantric deities in parts of the body. Both are done for the same reason: sanctification. Visualizing angels placed in the body transforms the body into a temple, healing the soul, and making it a suitable locale for prophetic insight.

With the legitimizing of new types of prophecy, prophetic ecstasy has returned to earth. New forms of interpretation can be accepted, and we have a canon which is no longer closed. Hebrew letters can be brought to life by breathing exercises and postures and can communicate new information. Through prophecy and holy inspiration (*ruach ha kodesh*) comes a cleaving to God, and a renewal for the religion and for the world. There are Ophanim centers in New York City and Jerusalem, as well as assorted congregations which have been influenced by their teachings.

According to another Renewal group, the Jewish Yoga Shabbat School, "prayer of the body" is a precursor to awareness of the unity of breath, mind, and spirit. They have developed the practice of "Torah Yoga," in which the Torah is like a body, with scriptures that have living law and wisdom. This form of Jewish yoga is performed on Shabbat, and letters of the Hebrew alphabet are used as mantras (much as they have been used in certain Kaballistic practices, with a focus on the words themselves rather than emotional or petitionary prayer). While traditional Kaballah restricted their teachings to men who were married and over forty years old, these restrictions are no longer valid for the more modern mystical and prophetic forms of Judaism. Indeed, one subtype is Rosh Chodesh Rav Yoga, which celebrates women's bodies and sisterhood, and follows the new moon rituals of the Bible.

Other subtypes of Jewish meditative ritual encourage other goals: Rav Yoga transforms mental and physical energy into spiritual energy, which is suitable for worship of God. Tefilla Yoga involves the visualization of an inner prayer service and the purification of the heart through prayer; there are matching yoga postures and sequences with themes and key words of traditional Jewish prayers.[38] The practice of Shabbasana ("where Shabbas meets vipassana") emphasizes the peace and rest of the Sabbath day and dissolves all differences and tensions, bringing a state of unity through stretching and meditation. These tensions return at the Havdallah service, when holy time ends and secular time begins again.[39] Meditation involves guided visualization, ayin consciousness, and nourishing the "soul stream." At their seva yoga seders, Jewish kirtan is performed.

The Jewish Renewal movement has been fueled by dissatisfaction with traditional Jewish belief and practice, which is understood as dull, unoriginal, and overly intellectual. As one rabbi states, she was dissatisfied with "dry, desiccated, repetitive Shabbat services," and felt betrayed and spiritually wounded by the resistance to spirituality and mystical experience that she found within the tradition.[40] Now, she has deeper experiences through chanting the names of God, guided visualization, and meditation on the River of Light and the inner centers of energy or Sephiroth. As Rabbi Schachter-Shalomi emphasizes in his book *First Steps to a New Jewish Spirit: Reb Zalman's Guide to Recapturing the Intimacy and Ecstasy in Your Relationship with God*,[41] modern Jews need to emphasize practice rather than theory, have compassion rather than authority, and perform meditation and contemplation.

There have been many critiques of the Renewal movement from more conservative groups within Judaism. Rabbi Neil Gillman of the Jewish Theological Seminary notes that some Jews are concerned about its emphasis on spiritual experience over observing Jewish law, and that its free-form style of worship waters down the tradition.[42] Its emphasis on continuing revelation evokes the same sort of criticism in Judaism that charismatics receive from fundamentalists in Protestant religion, who believe the canon to be closed. For those who understand Judaism to be a long history of interpretation and argument, leading to a more perfect understanding of the revealed word of God, the addition of Buddhist chants, Native American dances, and Sufi songs is a violation of the tradition. But for those interested in adaptation and assimilation, the adoption of new worship techniques makes Judaism modern as well as ancient, experiential as well as legalistic.

While Islam in the West tends to be associated with mosques and imams, there are ecstatic versions of Sufism that have become popular. Poetry books by Rumi, the Persian poet of the Mevlevi Order, have become modern best sellers, and groups of Whirling Dervishes have toured the world. But for modern Muslim ecstasy, we will include the Sufi Order of the West, and the Dances of Universal Peace begun by Samuel Lewis.

The Sufi Order of the West is a contemporary Sufi order which originated from the teachings of the Indian Sufi Hazrat Inayat Khan. Pir Vilayat Inayat Khan, the son of Inayat Khan, was a leader of the Sufi Order for over fifty years, until his death in 2004. He studied many religious and spiritual traditions and incorporated them into his teachings, adding to it the scholarship of the West in music, science, and psychology. He travelled widely, wrote books on Sufism, and founded

The Abode of the Message, a retreat center in New Lebanon that was formerly a Shaker site and is now also used by other Sufi orders and interfaith gatherings. This order was renamed Sufi Order Ināyati in 2016 and is currently led by Pir Zia Inayat Khan, the grandson of Inayat Khan and son of Pir Vilayat Inayat Khan. It teaches the essential unity of all faiths and accepts members of all creeds. Its Universal Worship Service honors the world's spiritual traditions with readings from the holy books of different religions. Students take classes in learning about the different religions and the spiritual essence of these traditions as part of their training and perform contemplative practices from many religions.

The traditional symbol of the Sufi Order of the West is the *Tughra Inayati* or the "Winged Heart," designed by the dervish Hafizullah in honor of Inayat Khan. The winged heart is an old Sufi symbol and was chosen by Inayat Khan as the seal of the "Sufi Order of the West" at its founding in 1910. The Arabic script of the wings (in mirror image of each other) reads: "*Ya Hazrati Inayat*," with *ya* meaning "O" (an invocation), and *hazrati* meaning "presence; a respectful title." It is the heart that flies into freedom.

The Sufi Order of the West has become associated with the Dances of Universal Peace. These dances were created by Samuel Lewis, popularly known as Sufi Sam or Sufi Ahmed Murad Chishti, who created the Sufi Ruhaniat Order. Murshid Samuel Lewis (1896–1971) was initially schooled and authorized as a teacher in a branch of the Chishti Sufi order, which was brought to Europe and the USA from India in 1910 by the Hazrat Inayat Khan. His teacher was Murshida Rabia Martin, who was initiated by Inayat Khan, and was designated spiritual head or Murshida of Sufism in America in 1927. Lewis himself also studied in India, Pakistan, and Egypt with a number of Sufi orders including the Naqshabandi, Kalandari, Qadri, and Rifa'i. The Dances of Universal Peace use movements derived from the practices of these orders as well as from a synthesis of spiritual dance techniques Lewis derived from his study with American sacred dance pioneer Ruth St. Denis.

Samuel Lewis became known in the late 1960s as the creator of the Dances of Universal Peace, sometimes called modern "Sufi Dancing." Lewis had studied many other traditions, but was most comfortable as a Western Sufi. His Sufi Ruhaniat International recognizes a spiritual lineage from Inayat Khan through Murshid Sufi Ahmed Murad Chishti. Lewis introduced modifications of traditional Sufi walking practices, which encouraged the ability to distinguish various states of awareness and control them. Teacher training in the Dances of Universal Peace includes training in walking meditation. In one such practice, the student is encouraged to walk in the footsteps of one's teacher or imagine the rhythmic presence of the teacher as one moves (a practice called *tassawuri*, or imagination).

While Samuel Lewis was also an ordained Rinzai Zen master and practiced many religious traditions personally, the strongest influence on the dances was Sufism. They involved chants, hymns, and songs from many religious traditions and brought people into ecstatic states during the dances. As their Web site notes, the dances were begun by Samuel Lewis in order to induce ecstatic experiences. "Murshid Sam, as he was known to his Sufi students in the 1960s, brought forward the Dances, meditative walks and other esoteric practices to awaken the hearts and minds of his Sufi students and create an ecstatic experience of unity."[43] These dances have spread around the world and are intended to create states of joy and peace in their participants:

> **The mood of the Dances is** infinitely variable, evoking feelings of love, joy, peace, compassion and ecstasy. The focus is on the sacred phrase, which often points to truths about Life which cannot be put into words. The dancers' experience flows out in the context of sacred presence to one other and to the group as a whole. An experience of the Unity of the human family with all of Life and a taste of one's essential nature are among the gifts the Dances bring to so many who come in contact with them.[44]

As one practitioner describes her experiences at the dances:

> I am an executive coach, and came to spiritual practice only very recently. Of the thousands of people with whom I have worked, at least one fifth approached me with questions about spirituality... I became involved in a 'spirit in the workplace' project, and later attended a mystery school where nine of the world's wisdom/traditions were studied. It was here I met a Sufi Master and was introduced to Body Prayer in the Dance Circle. I was inspired by this practice and joined the Dances of Universal Peace in my local area...There are two kinds of profound states that I find to be evoked by the Dances. One is a deep connection with the universe, a sense of personal sharing, in which prayer seems to journey through me out into the world as an offering. The other state I would call ecstasy. To me, this means existing in two worlds at once. In this state, the boundaries of my

> body disappear, and I seem to merge with all that is. I become a part of the cosmos yet I am still aware of the Dance; the movement, the song, and the other participants... In such situations, I feel as if I ride a wave of ecstasy created by others.[45]

The Dances of Universal Peace and its associated Walking Meditations are spiritual practices that involve motion. Drawing on the sacred phrases, scripture, and poetry of many religions, the Dances blend chant, live music, and evocative movement into a living experience of unity, peace, and integration. It is described as a "taste of our true nature" which opens to the possibility of a deep spiritual revolution within the person. Another practitioner gives details of her experiences during the dances:

I believe the Dances of Universal Peace evoke unique mystical dance experiences.... Almost every dance meeting, I experience any of these mystical states during slow trance dances:

1. I lose track of my body after the dance has been going for several minutes. I may then find my awareness is outside my body, surrounding it, like I am observing myself and others.
2. Paradoxically, I also often feel no sense of separation and deeply interconnected as if we are all one.
3. And, I feel as if I am in the center of the circle, without a body. My sense of space is very different. As if I am floating, suspended or spinning.
4. I lose track of time. A sense of timelessness or no time exists.
5. Positive emotions are heightened. Feelings of elevated and magnified joy, ecstasy, bliss, creativity, and inspiration.
6. Dreams are heightened for awhile after a dance meeting. I have had precognitive, lucid, problem-solving, creative, and inspirational dreams...

My being able to achieve spiritually inspired altered states regularly enriches every area of my life! I believe when I experience mystical states then others are benefited.[46]

The Dances of Universal Peace include music and dance from Christian, Jewish, Hindu, Buddhist, and indigenous religions, as well as Arabic chants from Islam. Lewis felt the dance could be presented to the public as a means to an expanded sense of the self through devotion as well as a form of "world peace through the arts." But Sufism as a mystical tradition has been rejected by more orthodox forms of Islam. Many Muslim groups do not accept the Dances of Universal Peace as Islamic, as their chants name deities other than Allah. Its rituals are too universalist for most forms of Islam. However, they have become popular for Western Sufis who are more open to a variety of spiritual practices.

All of these are responses to the lack of room for ecstatic states in the mainstream religions of the modern West. They show a religious backlash to the devaluing of individual experience in religions that have become concerned with orthodox belief and limitations on spontaneous behavior. Whether it is fear of the Devil or following tradition or simply inappropriate or disorderly before God, ecstatic states are forbidden in most conservative groups. Thus, we have charismatic forms of the religion branching out and creating new varieties to fulfill modern needs.

While we focus here on Western religion, we may note that Eastern religion has had a different set of issues in relation to the denial of ecstasy. It is not practitioners who deny its relevance, but rather Western scholars, who claim that ecstasy is a colonialist imposition upon native peoples.

Notes

1. The local inspirational Christian television station shown in Charleston, SC, is station INSP. Besides the syndicated evangelists discussing the imminent end of the world, they are selling shofars (in small and large sizes) to Christians who will sponsor Jews from Russia and Eastern Europe to come home to Israel. While ostensibly this is to help the Jewish people (who are portrayed as old and starving and incredibly grateful), this is also to help with Jesus' Second Coming, which requires the Jews to return to the Holy Land.
2. Oral Roberts, *The Drama of the End-Time* (Tulsa: Oral Roberts, 1963), p. 65.
3. For further discussion, see Richard Landes, *Heaven on Earth: The Varieties of the Millennial Experience* (Oxford University Press, 2011).
4. See the "Historical Account of the Rapture's Origin" at www.kingdomlife.com Catholic Theology has been much less open to millennial thought. It was condemned by Joseph Cardinal Ratzinger (who later became Pope Benedict XVI) in the 1995 Catechism of the Catholic Church: "The Antichrist's deception already begins to take shape in the world every time the claim is made to realize within history that

messianic hope which can only be realized beyond history through the eschatological judgment. The church has rejected even modified forms of this falsification of the kingdom to come under the name of millenarianism, especially the "intrinsically perverse" political form of a secular messianism." See *Catechism of the Catholic Church.* Imprimatur Potest, Joseph Cardinal Ratzinger. Doubleday, NY, 1995, p. 194.

5. Hal Lindsay, *The Rapture* (Toronto: Bantam Books, 1983), p. 169.
6. Ibid., p. 38.
7. Ibid., p. 40.
8. Ibid., p. 42.
9. Tim La Haye's book *Apollyon: The Destroyer Is Unleashed* has stories of people who continued to live after the Rapture, trying to kill themselves by jumping off buildings and putting their heads into ovens to avoid demonic locusts. It was published in 1999 by Tyndale House Publishers.
10. Tim LaHaye, *Rapture (Under Attack)* (Sisters: Multnomah Publishing, 1998), p. 23, as well as throughout the book.
11. Ibid., p. 138.
12. Marcus Bach, *The Inner Ecstasy: The Power and the Glory of Speaking in Tongues* (Nashville: Abingdon Press, 1969), pp. 15–17.
13. Ibid., pp. 178–179.
14. "Assemblies of God Beliefs," endorsed by the church's Commission on Doctrinal Purity and the Executive Presbytery, at www.ag.org/top/beliefs/baptism_hs/baptmhas_02_refilling.cfm.
15. Devereaux Jarratt, *A Brief Narrative of the Revival of Religion in Virginia, in a Letter to a Friend...*, cited in Vinson Synan, *The Holiness-Pentecostal Tradition: Charismatic Movements in the Twentieth Century* (Grand Rapids: William B. Eeerdmans Publishing Company, 1997), p. 9.
16. G. F. Taylor, *The Pentecostal Holiness Advocate*, cited in Synan, ibid., pp. 52–53.
17. Ibid., p. 12.
18. Frank Stagg, E. Glenn Hinson, and Wayne E. Oates, *Glossolalia: Tongue Speaking in Biblical, Historical and Psychological Perspective* (Nashville: Abingdon Press, 1967), p. 67.
19. Ibid., p. 100.
20. Ibid., p. 193.
21. Ibid., p. 213.
22. See Michael G. Moriarty, "The New Charismatics," in *Biblical Perspectives* IV, no. 3 (May–June 1991). Both of these latter groups have been controversial because of their authoritarian aspects.
23. Ibid., p. 275.
24. This was very controversial, with all sorts of accusations of deception and dishonesty and satanic activity. For some of the arguments, see http://deceptioninthechurch.com/pensacola.html.

25. William Darrow, "Modern Charismatic Confusion," Logos Resource Pages, at www.logosresourcepages.org/charismatic.htm.
26. Ibid., p. 15.
27. Kevin and Dorothy Ranaghan, *Catholic Pentecostals* (Paramus: Paulist Press Deus books, 1969), p. 131.
28. Cited in Ranaghan, ibid., p. 185.
29. Ibid., p. 192.
30. David L. Ariel, *Kabbalah, The Mystic Quest in Judaism* (Lanham: Rowman and Littlefield, 2005), p. xviii.
31. Ibid., p. xviii.
32. Ibid., p. xix.
33. See their website at www.aleph.org.
34. Ibid., from the Aleph website.
35. See the Ophanim website www.angelfire.com/pe/ophanim, for details.
36. Ibid., Ophanim website.
37. Ibid., Ophanim website.
38. For a 2017 advertisement, see https://patch.com/illinois/northbrook/calendar/event/20170624/152701/tefilla-yoga.
39. Information on this tradition may be found at www.rabbiheatheraltman.com.
40. As described by Rabbi Goldie Milgram at www.rebgoldie.com.
41. Rabbi Zalman Schachter-Shalomi, *First Steps to a New Jewish Spirit: Reb Zalman's Guide to Recapturing the Intimacy and Ecstasy in Your Relationship with God* (Woodstock: Jewish Lights Publishing, 2007).
42. Problems with the Jewish Renewal movement were discussed on the PBS television show "Religion and Ethics Newsweekly" for September 30, 2005. Some highlights from the discussion may be found at www.pbs.org/wnet/religionandethics/week905.
43. See http://www.dancesofuniversalpeace.org/about.shtm.
44. See Ibid.
45. Data from e-mail interview with Adrienne, 2006.
46. Data from e-mail interview with Debra, 2006.

CHAPTER 8

The Case of Hinduism: Ecstasy and Denial

While most writing on the topic of ecstatic religious experience has emphasized Western models, more recently Asian religions have been directly incorporated into the general arguments about religious experience. One critique which continues to be influential was written by Robert Sharf, in his article on "Experience" in Mark Taylor's *Critical Terms in Religious Studies*. He argues that:

1. Ideas of religious experience in Asia are not really indigenous ideas—they are "a relatively late and distinctively Western invention."
2. What earlier ideas exist in Asia about religious experience show that it is unimportant. There is no pre-colonial emphasis on experience, its importance only comes from Western-trained writers like Radhakrishnan. Religious authority is rarely based on "exalted spiritual states."
3. There are false, inconsistent or dubious claims about religious experience, such as claims of alien abduction. Since some claims of subjective religious experience are false, therefore all claims on the topic are false.[1]

In order to address these arguments against "exalted spiritual states" in the context of Hinduism, I shall look at terminology and specific ideas of religious experience in the major forms of Hinduism and discuss concepts of supernormal perception and spiritual discernment in Hindu

J. McDaniel, *Lost Ecstasy*, Interdisciplinary Approaches to the Study of Mysticism, https://doi.org/10.1007/978-3-319-92771-8_8

philosophy. To respond to the above arguments and generalizations, it is important to have specific counter-examples.

In Hindu tradition, there are several terms which fit different aspects of ecstatic religious experience. Many of these terms are quite ancient ones, not resulting from colonialism and "Western invention." Besides the various terms associated with *jnana*, such as *abhijnata*, translated as experience, knowledge, and wisdom,[2] there are two terms which fit more closely the English understanding of the term "ecstatic experience." One of these is *darsana*, and the other term is *bhava*.

Darsana is experience associated with seeing or observation, especially the direct perception of a deity by an observer. A relationship of mutual recognition between person and God is often implied by the term. It is used primarily to mean this form of religious vision, though the term can also refer to philosophical schools, and to "seeing" from a given perspective. Such *darsana* may be spontaneous, as when a statue of a God or goddess is suddenly perceived as living (and the observer is thus said to have *darsana* of the deity). Or the person may have spiritual knowledge from repeated direct perceptions (*bhuyodarsana*). Such experiential knowledge is considered to be very valuable in various types of Hinduism, especially in the bhakti traditions. As Diana Eck notes, the central act of Hindu worship, from the perspective of the lay person, is to stand in the presence of the deity, and to see and be seen by the deity.[3]

The Sanskrit term most widely used in West Bengal for ecstatic and mystical states, and for religious experience in general, is *bhava* (or in the Bengali vernacular, *bhab*). It is a term that has long been used in literature and in popular religion. The *Samsad Bengali–English Dictionary* gives such meanings as existence, condition, mental state, emotion, inner significance, reverie, mood, and ecstasy.[4] The Monier-Williams *Sanskrit Dictionary* has four full columns of definitions for the term *bhava*. The *Bangala Bhasar Abhidhan* has two columns, including the terms for essence, heart, divinity, yogic powers, rapture, and possession trance.[5] The term *bhava* has the sort of range and ambiguity that we find in the terms religious experience and ecstasy, and this range of understandings is found in both classical and vernacular sources.

I interviewed about forty Hindu informants on the meaning of the word *bhava* while doing research on local concepts of ecstatic states in West Bengal. They tended to divide the meaning of the term into secular (*laukika*) and religious or supernatural (*alaukika* or *adhyatmika*) definitions. Secular definitions include responses to art and beauty, emotion, passion, feelings, and ideas. Religious definitions include the experiences of holy men and women, the relationship between the soul (*jivatman*) and the God, surrender to the goddess, intuitive thought, forgetting the material world, and absorption in the deity. Such *bhavas* may be experienced inwardly, or acted out when the person is said to be *bhavavesa*, or overcome by *bhava*. Vaishnava literature also uses the term *anubhava*, which refers to ecstatic emotional states, and the ways that *bhavas* are expressed through the body. These terms represent valued states in Hinduism, from both the pan-Indian Sanskrit tradition and regional traditions of West Bengal. The popular use by laypeople of the term *bhava* for ecstatic state is usually a Bengali abbreviation for the term *mahabhava*, meaning great state or great emotion. This is a state of union with the deity or *brahman*, beyond *nama-rupa* (name and form). It is generally understood as beyond the symbols which are used to evoke and express it. As one Bengali Shakta informant phrased it,

> *Bhava* is very deep thought, deep in the heart, until one is lost within the self… There are three stages of *bhava* in the worship of Shakti- *bhava*, possession by *bhava* (*bhavavesa*), and deep trance (*bhava samadhi*). In *bhava*, one becomes lost in memory and emotion. In possession by *bhava*, one becomes lost from the material world and sees the heaven worlds. In the deepest state of *bhava*, one roams in the absolute (*atman*).[6]

India is most popularly known in the West for its yogis, but there are a wide range of ecstatic religious traditions there. In the Shaktism of West Bengal, there is a popular fusion of Advaita Vedanta and Bhakti Hinduism. It supports a sort of spontaneity that would be unacceptable in Western philosophy and Theology, for it accepts an Ultimate Reality or *brahman*, which may be shown in theistic (*saguna*) or non-theistic (*nirguna*) fashion. Because there is no single form in which *brahman* must be understood, no particular institution, text, or authority figure is decisive. As any particular name and form (*nama-rupa*) for a deity is considered to be illusory and secondary, religious schools, models, and categories are less important.

When I asked my Bengali informants for the most important sacred text of Bengali Shaktism, the text most quoted was the medieval *Kularnava Tantra*, which is quite suspicious of institutional knowledge and organization. It satirizes scholars or *pandits*, who emphasize the intellectual goal of tantric study and practice, and who value such terms as *vidya* (learning or scholarship) and *tattva* (essential nature, philosophical knowledge). The tantra itself emphasizes the importance of *pratyaksa* (immediate experience), *bhava* (mood, feeling, emotion, ecstatic state), *upalabdhi* (realization), and *ullasa* (blissful joy). It is one of the most important texts available for a popular understanding of ecstatic states. As it states,

> 87. O Beloved! Many ignorant people fall into the deep well of the six philosophies, but they are controlled by their instincts and cannot attain the highest knowledge.
> 88. They are drowning in the dreadful ocean of the Vedas and *sastras* (religious books), and they are driven in one direction and then another, by philosophical discussions and debates, which are like terrible waves and crocodiles.
> 89. (There are) people who have read the Vedas, Agamas, and Puranas, but who do not know the highest truth. All their knowledge is like the cawing of crows, and nothing more.
> 90. O Goddess, they turn their backs on truth, and read books day and night, always worrying about what they should be learning, saying this is knowledge or that is knowledge.
> 91. They know literary style, syntax, and poetry, and ways to make writing attractive, but they are fools, and they are confused and worried.
> 92. What they understand is not the highest truth, and what they interpret is not the real meaning of the sacred books.
> 93. They speak of ecstatic consciousness (*unmani-bhava*) but they do not experience it…
> 94. They chant the Vedas and fight among themselves, but they do not know the highest truth, as a cooking ladle does not know the taste of the things in it.[7]

This tantra (dated by scholars between 1000 and 1400 CE) tends to devalue traditional religious and textual authority, denying the modern constructivist approach through paradox. Following a particular religious tradition is not required for Bengali holy people. Many of India's "divine madmen" have no institutional affiliation, and if they do become part of institutions, their beliefs and ritual practices develop *after* the ecstasy, not before it. As the Kularnava Tantra notes, without having experience of ecstasy, the critic is like a ladle, unable to understand the soup it holds.

These terms for ecstasy represent highly valued states in Hinduism, from both the pan-Indian Sanskrit tradition and regional traditions of West Bengal. However, they are general terms, and in order to understand the development of the concept of ecstatic experience, it is useful to look at terms from within some major types of Hinduism.

Vedic Religion

In that notably pre-colonialist work, the Rig Veda, there are many descriptions of ecstatic religious experience. The Vedas are the oldest Hindu scriptures available to us, written in Sanskrit. The dating on the Rig Veda is much debated by scholars, with estimates ranging from 2000 BCE to 600 BCE. According to Gavin Flood, the Vedic collections or samhitas date to roughly 1700–1100 BCE, and the associated texts, as well as the editing of the samhitas, date to approximately 1000-500 BCE. Thus, we have a Vedic period, from the mid-second to mid-first millennium BCE, which would include the Late Bronze Age and the Iron Age.[8] In the Rig Veda, descriptions of ecstatic experience come from the *rishis* or seers, who describe visions of the gods and of other worlds, from the priests who take the drug *soma*, and from the long-haired ascetics or *kesins* who "ride the winds."

The Vedic seers had experiences often called mystical and ecstatic by scholars, a visionary "beholding" which allowed them to write inspired hymns with the words of the gods, and they describe the worlds of the gods or *devas.* Many hymns speak of the divine light that the *rishis* have seen and quote the words of the gods that they have heard. These hymns are considered by most forms of Hinduism as revelatory (*shruti*) and are highly valued. Ecstatic experience involves vision of the "shining ones," of the heaven of endless light (*svarga*), and of *brahman*, the source of the greatness of the gods.

The most famous description of ecstatic religious experiences is in the most important Veda, the Rig Veda. It is probably that of the drinkers of *soma*, whose insights were highly respected by the community. In Rig Veda VIII. 48, a seer speaks of his experiences:

1. I have tasted the sweet drink of life, knowing that it inspires good thoughts and joyous expansiveness to the extreme, that all the gods and mortals seek it together, calling it honey...

3. We have drunk the Soma, we have become immortal; we have gone to the light, we have found the gods…
5. The glorious drops that I have drunk set me free in wide space…
12. The drop that we have drunk has entered our hearts, an immortal inside mortals. O fathers, let us serve that Soma with the oblations [offerings] and abide in his mercy and kindness.[9]

The ritual can include a vision of the heavens, as Rig Veda IX.9 describes:

7. Where the inextinguishable light shines, the world where the sun was placed, in that immortal, unfading world, O Purifier, place me. O drop of Soma, flow for Indra…
11. Where there are joys and pleasures, gladness and delight, where the desires of desire are fulfilled, there make me immortal. O drop of Soma, flow for Indra.[10]

Another example of ecstatic experience from Vedic times was the Kesin hymn of Rig Veda X.136, which describes the flight of the long-haired ascetics who were believed to visit other worlds and become possessed by gods. The hymn moves from third person to first person:

2. These ascetics, swathed in wind, put dirty red rags on. When gods enter them, they ride with the rush of the wind.
3. "Crazy with asceticism, we have mounted the wind. Our bodies are all you mere mortals can see."[11]

While it is a more common style to write in the third person for hymns, there are many accounts which speak directly of personal experience, and in them the seers speak the words of the gods directly. The Vedic seer perceives through *dhi*, inner vision, until the inner sun is like the fire God Agni in his heart (and most Vedic seers were male). It is important for the seers to describe such visions, for according to Jan Gonda, they are incomplete until they are described and come to have lasting value only when they can evoke the original vision.[12]

The Vedic seers describe the light of lights, which is sweet as honey, and its overflowing sweetness brings human beings to the land of immortality. They go to the gods, ascend into the light, and they are transformed. Ecstatic religious experience is important in Vedic tradition, because it gives both supernatural vision and immortality. It is understood to connect practitioners with the source of the universe.

Upanishadic Religion

In the Upanishads, which are commentaries on the Vedic texts, we have many stories and metaphors which describe the practice and goals of religious contemplation. Over two hundred Upanishads are known, with the first dozen as the oldest and most important ones. These are found toward the end of the Brahmana and Aranyaka texts and initially were memorized and passed down orally. Their dating is much debated; Patrick Olivelle suggests that almost half of these early Upanishads are pre-Buddhist.[13]

These texts contain first-person accounts of contemplative states, which in general involve blissful joy, ecstatic union with ultimate awareness or *brahman*, and the dissolution of the individual ego. Brahman is the Absolute, ultimate Being, that which is truly Real. Experience of *brahman* is compared to a grain of salt dissolving in water and becoming one with it, and with the rivers that run to the sea and enter it fully. A good description comes at the end of the Kundika Upanisad:

> Stirred by the wind of illusion, the waves of the whole universe
> Repeatedly rise and fall within me, the ocean of total bliss…
> Like the sky am I, far, beyond the reach of time…
> Like the sea am I, without a farther shore…
> I am pure consciousness, the witness of all!
> I am free from the thought of "I" and "mine"! I have no lord!…
> I do not act, I do not change, I have no parts, I have no form.
> I am eternal, I have no thought, I am unique, I have no support.
> All are myself and I am all! I am unique and I transcend all!
> I am my own eternal bliss, pure undivided consciousness![14]

In the Paramahamsa Upanishad 2.49, the state of *brahman* is described more simply:

> I am indeed that calm and unchanging Being, a single mass of bliss and consciousness. That alone is my highest abode. By knowing that the highest Self and the lower self are one, the difference between them dissolves into oneness.[15]

Sometimes, *brahman* is understood as a force which grasps the individual, and philosophy and ritual cannot determine its actions. As the Mundaka Upanishad 3.2.3 states,

> This self cannot be grasped by teachings or by intelligence, or even by great learning
> Only the man he chooses can grasp him whose body this self chooses as his own.[16]

It is direct experience which gives knowledge of brahman, as the Kena Upanishad 2.4 states:

> When one awakens to know it one envisions it, for then one gains the immortal state. One gains power by one's self (*atman*). And by knowledge, the immortal state.[17]

This value of this state of experience is described by the sage Yajnavalkya in the Brihadaranyaka Upanishad 4.3.32–33. As he explains to the king:

> He becomes the one ocean, he becomes the sole seer! This, Your majesty, is the world of *brahman*." So did Yajnavalkya instruct him. "This is his highest goal! This is his highest attainment! This is his highest world! This is his highest bliss! On just a fraction of this bliss do other creatures live.[18]

The Upanishads speak of the divine eye with which people perceive *brahman*, which is the light of lights. Union with *brahman* is supreme unsurpassable bliss, indeed there is a level of the self or *atman* composed entirely of bliss. This union brings freedom from death and rebirth, from suffering, from ignorance, from "the knots of the heart," desire, doubt, and fear. Such experience is eternal delight and peace, and unlimited freedom in all worlds. One is able to recognize truth, consciousness, and bliss (*sat, chit, ananda*), and alternatively goodness, beauty, and truth (*sivam, sundaram, satyam*).

While the Upanishads are often accused of avoiding descriptions of religious experience, and being suspicious of claims of personal experience, if we look closely we can see that the religious claims of these texts often come from first-person accounts. It is perhaps paradoxical to say that the ultimate personal religious experience is merging with the impersonal *brahman*, but this is the perspective of the Advaita Vedanta school. Though the writing involves paradox, as does most mystical writing, we nevertheless see writers who are trying to describe ecstatic experiences which they find difficult to express in ordinary language.

The Yoga Tradition

There are many understandings of yoga in Hinduism, for the term "yoga" means both union and discipline. Concepts of union and practices of discipline are widespread in Hindu subgroups. Perhaps the form of yoga most well-known in the West is the classical yoga described by the sage Patanjali. According to the Yoga Sutra of Patanjali, liberation comes through asceticism and contemplation. One of the most important experiences in meditation is *citta-vritti-nirodha*, the silencing of the activities of consciousness. This leads to the state of perfect contemplation or *samadhi*, which in turn leads to the ultimate state of liberation or *kaivalya*. During meditation, consciousness becomes like transparent crystal, and eventually the person may become the object of concentration, "shining with the light of the object alone." While liberation is the ultimate goal, the Yoga Sutra details the various supernatural experiences or *siddhis* which may occur during certain types of meditation. These perfections or attainments may occur due to birth (and the karma from past lives), drugs, mantras, *tapas* (asceticism and purification), and states of *samadhi*.[19] They allow the person to develop discrimination and to able to differentiate between pure spirit (*purusa*) and the *gunas* or aspects of life and thought which belong to matter, and they are aids in the development of concentration. They are less valued than the higher forms of experience described in the Yoga Sutra I, 18–19, which involve the two forms of *samadhi*: These are *samprajnata* (achieved by yogic effort, and which has four substages) and *asamprajnata* (which occurs spontaneously, to both humans and supernatural beings). Religious ecstasy which is induced by ritual and ecstasy which occurs spontaneously are recognized within a wide variety of Hindu religious texts.

Other texts on yoga emphasize the dissolution of the intellectually created world of concepts, and the ecstatic revelation of the interior world of the spirit. Sense perception is suspended, and the person experiences the bliss of dissolution in the practice of *laya yoga*. The *Hatha Yoga Pradipika* describes the yogi as "empty within and without like an empty pot in space, and also filled within and without like a pot in the ocean."[20] In the ultimate state, the yogi is empty yet full, experiencing the union of *jivatman* (individual soul) and *paramatman* (Absolute or highest Self). All of these experiences are important as developmental stages, but the highest stages are the most valued religious experiences.

While most yogic texts are written in the form of handbooks and manuals, we do have personal accounts of yogis and yoginis. For instance, the Kashmiri yogini Lalla Devi describes her meditation (in poem 31):

> I pulled the reins of the steed of the mind
> I compressed the life force circulating through the ten channels…
> Concentrating on the om-sound
> I made my body like blazing coal.
> Leaving behind the six crossroads
> I traveled the path of Truth.
> And then I Lalla, reached the Abode of Light.[21]

The major value of the yogic meditation is to lead to the ultimate ecstatic experience, to reach the abode of infinite light, and thus to gain perfect liberation (*jivanmukti*) in this life.

The Tantric Tradition

There is a close relationship between the yogic and tantric traditions. The origins of tantra are usually said to be around 500 CE (though some scholars point to older precursors).[22] Over the next thousand years, it elaborated theories and practices, which have influenced both philosophical and devotional traditions in Hinduism. The tantras usually emphasize the gods Shiva (especially in the Kashmiri Kaula tradition), or Krishna (in the Vaishnava Sahajiya tradition), or the goddess Shakti (in the South Indian Shri Vidya and Bengali Shakta tantric traditions). Tantric ideas and rituals have also been influential in Buddhism, especially Tibetan Vajrayana.

While Hindu bhakti or devotionalism describes the highest religious experience as intense love of the deity, in tantra it is union with a deity that is considered not only possible but ultimately the goal of human life. Tantra emphasizes the union of opposites: spirit and matter, *brahman* and the physical world (*jagat*), and God and humanity. While the modern sensationalizing of tantra has emphasized sexual union, this is only one form of union among many others which are more important to the tradition. In Kaula thought, the highest goal is the "vision of the truth of the *kula* (*kula-tattva-artha-darsana*)" which is the direct intuition of Ultimate Reality. One gains the God Shiva's perception, in which all things appear the same, divine and otherwise (*sama-darsana*). In the Shri Vidya tradition, the earlier religious experiences, the momentary ecstasies, or states of *kshana samadhi* are temporary instances of the highest state of *sahaja samadhi*, a spontaneous ecstatic state in which the person can perceive any world, divine or human.

Many tantric texts discuss the importance of ecstatic religious experience. In some, ecstatic states act to verify religious claims. In others, these states are themselves the goal of the practice. The medieval *Kularnava Tantra* has long been an important text for both Kashmiri and Bengali tantric traditions and was a major text used by modern informants which I interviewed in West Bengal. It describes the state, and the value, of absorption in the God Shiva, which it calls *samadhi*.

> 9.14 He [the yogi] does not hear, or smell, or touch, or see; he does not know pleasure and pain, he does not analyze. Like a log, he does not think, he is not aware of anything [material]. One who is thus absorbed only in Shiva is said to be in *samadhi*.
> 9.15 Just as no differences exist when water is poured into water, and milk into milk, ghee into ghee, so there exists no difference between the individual self (*jivatman*) and the highest self (*paramatman*)...
> 9.25 In comparison to the pure and supreme state of consciousness attained by the great yogi, even the states of gods and other divine beings have no value.
> 9.26 For one who has seen (*darsana*) the all-pervading, peaceful, blissful and imperishable, nothing remains to be attained or known.[23]

The *Kularnava Tantra* is a text which speaks clearly and unambiguously about the importance of experience, especially that of union with Shiva (*sivatva*):

> 8.85. He gazes outward but looks inward, and his eyes are unblinking. This is the *sambhavi mudra*.
> 8.86. This is the greatest of *mudras*, the true form of the bliss of union (*samarasa*), expressing the nature of the self. By means

of this, the devotee truly becomes Shiva. There is no doubt about this.

8.87. Persons engaged in study of the self can know some of this bliss. But this state is beyond description and must be experienced, as the pleasure of drinking sugared milk must be experienced.[24]

The tantric tradition involves complex ritual actions, and these actions (*sadhana*) engender a wide variety of ecstatic religious experiences, which are described in the texts. The *Kularnava Tantra* gives one set of examples.

The Dharma Tradition

Of all types of Hinduism, this tradition has the least emphasis upon ecstatic religious experience. Leading the dharmic life is fulfilling one's obligations according to age, gender, caste, region, and many other factors. The goal is being successful, living well as a householder if one is an adult, supporting past and coming generations, and living a moral life. In the past, members of this approach to Hinduism would contribute to temples and give offerings to deities, but no personal experience was expected. This tradition has been much influenced by modern secularism and Westernization, and some members today deny the existence and value of religious experience. Others may put off religion for a future rebirth, or at least until retirement, at which time they may withdraw from the world and follow a guru or perform meditation and other religious practices. There is a long tradition of tension in India between dharma (the good life, lived in harmony with the universe) and *moksha* or liberation, which leaves the world behind. It is not difficult to argue for the unimportance of ecstatic religious experience in Hinduism if this is the only type of Hinduism that is examined.

Folk Religion

Folk and tribal religions are likely the oldest continuous forms of religion in India, but as they are largely based on oral tradition, it is difficult to prove their origins and examine their development. There are debates whether folk and tribal religions are subtypes of Hinduism, or separate belief systems—though clearly many types of folk religion incorporate Hindu deities and worship rituals.

There is no one definitive type of folk or tribal religion in India, but we can generalize a few common themes. Possession by gods and local spirits tends to be very important, and people who experience trance states are often understood as lay religious authorities. Sometimes, there is group possession by ancestors as well, which is usually understood as a message for the community (in West Bengal, trance is often sent by the primordial ancestress, the old woman or *budi ma*). Revelatory dreams and dream commands by deities are valued, and in many cases these are the origin stories for pilgrimage sites and holy places. These places are often natural locales where deities are believed to live, or burial grounds where ancestors dwell. There are religious specialists, male and female, called by various names (in Bengali, *ojhas* and *gunins*), who undergo training in healing and other rituals. Often their training involves images of death and ascetic practices and is accompanied by visionary experiences and vivid dreams. While sometimes such roles are hereditary, more often they come as a result of a "call" from a deity or an unusual experience that the person cannot explain otherwise. Many folk practitioners that I have interviewed were called in this way.[25]

Scholars have debated whether Bengali folk religion can be characterized as shamanic. While it occasionally involves death and rebirth experiences and flight of the soul, its most shamanic aspects would be possession trance, communication with supernatural beings (local gods and ancestors), healing by mantra and empowered objects, and community recognition of practitioners as religious authorities.

The Bhakti Tradition

The majority of Hinduism today involves some form of *bhakti*, or loving devotion to one or more deities.[26] This devotion should not merely be ordinary respect and obedience, but *parama prema*, the highest love, which brings a person to perfection. It is passionate longing for God's presence, and the joy which results from that longing (*premananda*, the bliss of selfless love) brings both immortality and knowledge of the God. The *bhakti* tradition rejected religious limitations on the basis of caste and gender and offered access to God to a wide range of people.

There are many *bhakti* denominations or lineages (*sampradayas*), with the largest number of devotees worshipping the God Vishnu and his avatars or incarnations (including Krishna and Rama). These are followed by dedication to the God Shiva and to the goddess Shakti or Devi (especially known in her forms of Kali, Durga, and Parvati).

There is Virasaiva poetry from the twelfth century which shows the role of ecstatic love in its poet–saints. One of the most famous is Basavanna, who worshipped the God Shiva is his form of Lord of the Meeting Rivers. As he wrote,

> When
> Like a hailstone crystal
> Like a waxwork image
> The flesh melts in pleasure
>   How can I tell you?
> The waters of joy
> Broke the banks
> And ran out of my eyes.
>
> I touched and joined
> My lord of the meeting rivers
> How can I talk to anyone of that?[27]

Another poet–saint is the twelfth-century poetess Mahadeviyakka, who called Shiva the lord white as jasmine. She was in love with him, though forced to marry a human husband. She wrote:

> O mother I burned
> In a flameless fire
> O mother I suffered
> A bloodless wound
> Mother I tossed
> Without a pleasure
> Loving my lord white as jasmine
> I wandered through unlikely worlds.[28]

While *bhakti* can refer to a variety of approaches to the deity, the most intensely loving form is called by Friedhelm Hardy "emotional Krishna bhakti" (as distinguished from the earlier tradition of *bhakti as* loyalty and respect). Hardy writes that it grew up with the Bhagavata Purana and the Alvar saints of South India, who wrote in the seventh to ninth centuries CE.[29]

Since that time, many Vaishnava works have discussed the importance of religious experience. In his tenth century CE *Bhakti Sutras*, which are dedicated to the God Krishna, the writer Narada describes degrees of devotional love. One begins by glorifying and appreciating the God's greatness, then loving his beauty, worshipping him, and remembering him constantly, identifying with being the God's slave, then his friend, then his parent, and then loving him as a wife loves her husband. The devotee should entirely surrender to Krishna and feel absorbed in him, yet still feel sorrow at the pain of separation from him. This feeling of union yet separateness is considered to be the highest religious state in Sutra 82, as described by Narada.[30] Such experience is not intended to be left behind when a person achieves liberation—it should ideally go on forever.

The *bhakti* writer Ramanuja identifies the God Vishnu with the *brahman* of the Upanishads, and Vishnu comes down to earth from heaven to give mankind salvation. The most important religious event is self-surrender or *prapatti*. This is a state of absolute delight, resulting from Vishnu's mercy, and the highest state to be attained by human beings. For the *bhakti* writer Madhva, the ideal state for the devotee is reflecting God's splendor, perfectly identifying with Vishnu, and seeing the world through his eyes. For the writer Vallabha, God's grace leads the devotee to eternal passionate love, in which can participate in the God's eternal play (*nitya lila*). The devotee attains the emotional states of *bhajananda* (the bliss of love) and *svarupananda* (the bliss of perceiving the God's true form) and becomes like Krishna's milkmaids or *gopis* with their minds entirely focused on the God.

One of the clearest examples of the importance of religious experience in Hinduism comes in the Gaudiya Vaishnava tradition of West Bengal. This tradition was inspired by the fifteenth-century saint Caitanya Mahaprabhu, who was believed to be a joint incarnation of the God Krishna and his consort Radha. The highest religious goal of this tradition is to experience all of the variations of intense love, the sort of passionate love experienced by Radha and her milkmaid friends the *gopis*. The legitimacy of these states of love is shown by the ecstatic experiences known as the *sattvika bhavas*, which include trembling, sweating, paralysis, crying, hair standing on end, changing skin color, and loss of consciousness. Thus, we see that religious experiences of intense love are not only legitimized by institutions, they may also be legitimized by other religious experiences.

Such emotional states may be results of meditative practice in current or past life, or occur spontaneously as a gift of Krishna's grace. When the states of intense emotion (*bhakti bhava*) develop, the emotion is said to be deepened and condensed, with a heart full of continual burning desire. Separation from the God is misery, while his vision gives infinite pleasure. The devotee goes through many stages of love until he or she reaches the highest state or *mahabhava*, which has two modes: sharing the loving bliss found only in Radha and the *gopis*, and the sharing the state of passionate delirium found only in Radha herself, in which she experiences all possible emotions simultaneously. Texts like the *Bhaktirasamrtasindhu* and the *Ujjvalanilamani* describe the spiritual development of the devotee, from the mild emotions which occur as a result of ritual to the passionate love which comes from true devotion.[31]

The religious goal of most forms of Gaudiya Vaishnava ritual practice is ecstasy—to bring the person a direct vision of the God Krishna and to have a continually new and passionate relationship with him. This religious experience is not tangential, it is central. The goal of the practice is not eliminating impurity or gaining power. It is to have the vision of Krishna forever.

For some forms of Gaudiya tradition, the ability to love Krishna involves developing a new spiritual body, known as the *siddha deha*. This is an eternal body, in the form of a handmaiden to the *gopis*, and it is understood as immortal. The religious emotions engendered by intense Vaishnava practice are too strong for an ordinary soul to bear—one needs a special spiritual self whose essence is love. This self holds the religious passions in this life and continues to be in the presence of Radha and Krishna in their eternal paradise after death. Because the flow of Krishna's love is like an ocean of bliss, a body of non-physical bliss is needed to contain it.

While this is the highest state in Gaudiya Vaishnava Theology, we do see other states that are valued: the appreciation of beauty and its essence (*rasa*), the joy which comes through chanting mantras and singing hymns (*kirtan*), and the appreciation which comes during the visualizations of the Vaishnava paradise. While the yogic traditions emphasize the importance of religious knowledge, Vaishnavism values love more highly, and the true devotee is one who experiences and expresses that love.

Arguments

Let us take another look at the claims about religious experience in Asia that were mentioned at the beginning of this article. Is it the case that there was no indigenous value on religious experience, and that it was simply lacking in Hindu religious literature until the British educational system put it there? Clearly not—postcolonial writers did not invent the importance of states like *brahman, kailvalya, samadhi*, and *mahabhava*. Nor can we say that interpretations that value such experiences are recent—there are long commentarial traditions that analyze and encourage such experiences, from the Upanishads onward.

Can we say that the idea of religious experience itself is so vague that it is useless? This might be the case for analytic philosophy, or for pragmatism, where the only worthwhile questions have very clear and unambiguous answers. However, such an approach denigrates all subjective experience, so that appreciation of music, the feelings of love and joy, and emotional involvement with drama and theater are all useless to discuss—and the Hindu *rasa* tradition of aesthetic experience values all of these.

As we compare understandings of religious experience in Hinduism and in the Judeo-Christian West, what is striking is their similarity. Both differentiate between natural and supernatural religious experiences, both have understandings of divinity as personal (a God or gods) and impersonal (an Ultimate Reality or Source), and both have types that value religious love and wisdom. While the West may have more emphasis on conversion and faith, and India more emphasis on mystical union and purification, there is a large area of overlap between their understandings.

Is it the case that all religious experience can be dismissed because of false claims about it by individuals? In Christianity, there is a long tradition of spiritual discernment, the ways to distinguish the origins of religious claims in God, Satan, or the human imagination. There are many reasons for false claims—voluntary or involuntary, originating from fantasy or trauma, from misunderstanding or sensory error. In the West, this is why spiritual direction has long been important to religious practitioners.

In Hinduism, we see the “discernment of spirits” in both the role of the guru and the philosophy of religion. The guru is a person whose authority is quite often charismatic and usually based upon those "exalted spiritual states" mentioned as irrelevant by Sharf at the beginning of this chapter. However, it is these states which attract followers and allow the guru to be understood as able to evaluate the states of others. While some gurus gain their status from heredity and lineage (there are people who are born into guru status, and others who are named by institutions), in West Bengal most gurus gain their role based on their experiences, the *bhavas* which they undergo and which they can transmit to others. Guru and saint (*siddha*) biographies abound in stories of spontaneous supernatural events, religious emotions shared by crowds in "waves of *bhava*," gurus who have visions and can read the thoughts of their devotees and bless them in special ways. Many gurus are outside of lineages and institutions, with only an initiating figure far in the past (whose religious persuasions are often unknown), and sometimes only a call from a God or goddess in a dream or vision. Yet they are understood to be able to evaluate and guide the religious experiences of others and are often called upon to do so.

The questions of false or illusory experience, whether alien abductions or sky flowers or ropes mistaken for snakes, have long been of interest to Indian philosophy. Hindu philosophers did not make the blanket claim that all experiences were reliable, but they also did not dismiss all experiences because some claims were problematic. Many thinkers would allow for the existence of a "flash of intuition" (*pratibhajnana*) about a future event. More controversial were the intuitions of sages (*arsajana*) and yogis (*yogipratyaksa*). For sages, the merit which resulted from austerities was understood to create a special type of perception, knowledge which is valid but not of sensory origin. Yogic perception comes about through the removal of mental impurities and can include supernatural perception (*siddhadarsana*). Yogic perception has often been divided into two types: ecstatic, which gives insight into the essential nature of the universe, and non-ecstatic, which gives information of subtle, hidden, and remote objects.[32] Because yoga philosophy traditionally accepts that all events, past and future, exist simultaneously, and that temporal order is a construction of the intellect, the yogi may have access to all times and all objects.

However, there is room for error. The yogi's perceptions may be distorted by such qualities as illusion, egotism, and restlessness. A major problem is the identification of self and mind, which can appear identical but are different. The mind (*buddhi*) is changeable and capable of error, subject to past memories (*samskaras*) and their intellectual and emotional associations, while the deeper self is beyond these problems. There may be misidentification, distorted interpretation, or simply wrong perception. There are supernatural events which precede the state of liberation, and they can be confused with illusory ones. It is only when the sage is understood to be free of the effects of karma that his or her experiences become true for all devotees. Otherwise, they are judged in terms of probabilities and individual faith in the guru and God.

We also have a school of thought on Indian religious experience which admits its existence, but devalues it and understands it be a sign of illness. Since Jeffrey Moussaieff Masson's writing about India's religious mysticism as a set of pathological states, with saints and ecstatics as traumatized individuals whose mystical visions reveal a cultural type of "psychic fainting," Asia in general and India in particular have been interpreted as a place of world-denying people who dread experience and long for escape from it.[33] Masson talked about the flight from emotion in Indian religion, and this view has become popular among some later scholars (with psychoanalytic training and otherwise). Ecstatic states are flights from trauma, ways of avoiding "real" life, much as Freud interpreted all inner experience as a flight from reality and a focus on Thanatos or the death urge. From this perspective, religious ecstasy is admitted as significant, but it is interpreted as pathological or least regressive.

Hindu practitioners in India today have varied ways of understanding mystical experience. During fieldwork in India, I interviewed renunciants, brahmins, people subject to possession trance, yogis, and believers of many sorts. Most did not speak English. In the modern West, the constructivist approach (the assumption that all religious imagery is socially constructed) is often used to understand world religious experience. However, in West Bengal, the only people who agreed with the Western categories of constructivism described were the Communists (members of the CPIM) with whom I spoke—and many of them were "secret practitioners" (they came to the burning ground after midnight, when nobody could see that they were meditating). Their view of history would accept religion as being historically constructed. And as Communists, their views were not Hindu, but rather Western.

I discussed constructivist ideas, that each culture creates its own gods, with some of the renunciants interviewed—I was curious how they would respond to them. Constructivism did not make sense to any of the renunciants interviewed, except perhaps for visualizations in specific *sadhanas* (rituals). People did not have the ability to create the forms of gods, and gods could take whatever bodies they chose. They were not limited to past incarnations. While it is helpful for experiences to fit into the known standards for particular groups, many sadhus had experiences which did not fit into the existing religious expectations. Many practitioners interviewed spent years going from one guru to another, attempting to understand their own spontaneous experiences, which often did not fit into traditional models. They were not "perennialists," not philosophers and linguists, but they had experiences which they felt were important and they sought to understand them. They also sought to learn the appropriate rituals to perform, an important sign of the acceptance of an idea in India. These seekers—who were native Hindu, not Westerners—might find a deity to worship, or a yogic meditation to follow, or an ashram to join. All of these were attempts to seek a place in the religious world when their experiences did not fit the accepted model.[34] They got to be accepted when they gained disciples and had some rituals to teach.

Academics in India could understand the idea that people might imagine religious experiences to gain power and respect. But for the more serious renunciants (sadhus and sadhikas), this was hypocrisy or insanity. A *bhava* was a way to understand the *alaukika* or supernatural worlds, but they could only be really known after a person entered them. Understandings given by the ancestors are sacred, but there is always more new to learn from an infinite ocean of possibility. As one sadhika (holy woman) phrased it, our world is small, but the universe is large—we should never claim to know everything, especially about the gods.

If we return to the critiques that began this paper—that religious experience is not an indigenous idea, that it is unimportant, and that all claims on the topic are false—we can see that both Indian religious history and Indian practitioners would have problems with them. Indeed, there are Indian scholars today who are speaking against these ideas, saying that constructivist and colonialist writers are stealing away their religious beliefs and values by taking credit for them, and claiming to know what they "really" believe. It is a case of cultural theft.

This chapter is an overview of some ideas on mystical and ecstatic experiences in Indian thought that did not come with colonialism, but existed long before the Europeans ever came to visit. Indeed, Indian ideas on ecstasy influenced Western thought in many ways—William James was influenced by the ideas of Vivekananda, and the

Transcendentalists came to appreciate the mysticism of the Upanishads. But it argues against religious ecstasy as a "relatively late and distinctively Western invention."

The study of ecstasy has been important in the history of religions—the next chapter will go over some ideas of major writers who analyzed the idea and include more modern analyses of ecstasy as a special form of consciousness.

Notes

1. Robert Sharf, "Experience" in *Critical Terms in Religious Studies*, ed. Mark C. Taylor (Chicago: University of Chicago Press, 1998). It has since been reprinted in Craig Martin and Russell McCutcheon, ed. *Religious Experience: A Reader* (Bristol: Equinox Books, 2012).
2. Sailendra Biswas and Birendramohan Dasgupta, ed., *Samsad Bengali–English Dictionary* (Calcutta: Shishu Sahitya Samsad, 1983).
3. Diana Eck, 1981.
4. *Samsad Bengali–English Dictionary*, ed. Sailendra Biswas and Subodhchandra Sengupta (Calcutta: Sahitya Samsad, 1979).
5. Jnanendramohan Dasa, *Bangala Bhasar Abhidhan* (Calcutta: Sahitya Samsad, 1979).
6. June McDaniel, *The Madness of the Saints: Ecstatic Religion in Bengal* (Chicago: University of Chicago Press, 1989), p. 24.
7. *Kularnava Tantram: Mula, Tika O Banganubadasaha*, ed. Upendrakumar Das (Calcutta: Nababharata Publishers, 1363 B.S.). This is a Sanskrit text, with a Bengali translation and commentary. I, 87–94. The translation here is my own, from the Bengali.
8. Gavin Flood, *An Introduction to Hinduism* (Cambridge: Cambridge University Press, 1996).
9. Wendy Doniger O'Flaherty, *The Rig Veda: An Anthology, 108 Hymns Translated from the Sanskrit* (Harmondsworth: Penguin Classics, 1981), pp. 134–135.
10. Ibid., pp. 133–134.
11. Ibid.
12. Jan Gonda, cited in N. Ross Reat, *Origins of Indian Psychology* (Berkeley: Asian Humanities Press, 1990), p. 106.
13. Patrick Olivelle, *The Early Upanishads* (Oxford: Oxford University Press, 2014), pp. 12–14.
14. Patrick Olivelle, *The Samnyasa Upanisads* (Oxford: Oxford University Press, 1992).
15. Ibid.

16. Ibid., 1996.
17. Ibid.
18. Ibid.
19. Rama Prasada, trans. *Patanjali's Yoga Sutras, with the commentary of Vyasa and the gloss of Vachaspati Misra* (New Delhi: Munshiram Manoharlal Publishers, 1988).
20. *Hatha Yoga Pradipika*, trans. Hamza Ali par. 45.
21. B. N. Parimoo, *The Ascent of Self: A Reinterpretation of the Mystical Poetry of Lalla-Ded* (Delhi: Motilal Banarsidass, Delhi, 1978).
22. For a good discussion of the issues, see Gavin Flood, *The Tantric Body, The Secret Tradition of Hindu Religion* (London: I.B. Taurus, 2006).
23. *Kularnava Tantram: Mula, Tika O Banganubadasaha*, ed. Upendrakumar Das (Calcutta: Nababharata Pablisars, 1363 B.S.), II. 88–90. This is a Sanskrit text, with a Bengali translation and commentary. The translations here are my own.
24. Ibid., VIII. 85–87.
25. See McDaniel, *Offering Flowers, Feeding Skulls*, Chapter 1 on Folk Shaktism.
26. Klaus K. Klostermaier, *A Survey of Hinduism* (Albany: State University of New York Press, 1994).
27. A. K. Ramanujan, trans. *Speaking of Siva* (New York: Penguin Books, 1985), p. 89.
28. Ibid., p. 121.
29. Friedhelm Hardy, *Viraha-Bhakti: The Early History of Krisna Devotion in South India* (Oxford, New York and Delhi: Oxford University Press, 1983).
30. Swami Tyagisananda, *Aphorisms on The Gospel of Divine Love or Narada Bhakti Sutras* (Madras: Sri Ramakrishna Math, 1978).
31. June McDaniel, *The Madness of the Saints: Ecstatic Religion in Bengal* (Chicago: University of Chicago Press, 1989).
32. Jadunath Sinha, *Indian Epistemology of Perception* (Calcutta: Jadunath Sinha Foundation, 1969).
33. Jeffrey Moussaief Masson, *The Oceanic Feeling: The Origins of Religious Sentiment in India* (Dordrecht: D. Reidel, 1981).
34. For a discussion of spontaneous ecstatic states in West Bengal that did not fit well into existing religious models, see McDaniel, ibid., 1989.

CHAPTER 9

Ecstasy and Empathy: Some Venerable Elders and New Directions

While many modern writers in Religious Studies are more interested in politics, economics, and redefining the term "religion" out of existence, we may remember that there have been writers more empathic toward religious ecstasy in the past. There are also some useful new directions in the present. But for the past, we may note three major emphases in the study of religious ecstasy: the analysis of emotion through phenomenology (seen in Rudolf Otto), the analysis of mind through psychology (seen in William James and Carl Jung), and the analysis of being through ontology (in Paul Tillich and Mircea Eliade).

Rudolf Otto (1869–1937) explored the role of emotion in religious ecstasy. In his book *Das Heilige*,[1] he describes states of ecstasy as "numinous" experiences. For Otto, the numinous state involves rapture, bliss, and "creature-feeling," when the person is overwhelmed by the "supreme power of the divine." This state may be experienced as mysterium tremendum, when the numinous is known primarily by intense and passionate emotion, or as the fascinosum or aspect of divinity that is blissful. As Otto describes the tremendum:

> The feeling of it may at times come sweeping like a gentle tide… it may burst in sudden eruption up from the depths of the soul with spasms and convulsions, or lead to the strangest excitements, to intoxicated frenzy, to transport, and to ecstasy.[2]

J. McDaniel, *Lost Ecstasy*, Interdisciplinary Approaches to the Study of Mysticism, https://doi.org/10.1007/978-3-319-92771-8_9

This state reveals a reality which is hidden, extraordinary, and beyond human understanding; it "completely eludes apprehension in terms of concepts."[3] It may be marked by joy, or by fear and dread and trembling. Both are aspects of ecstatic experience for Otto. There may be the recognition of the "overpoweringness" of the religious object for the person, or identification of the individual personality with the Transcendent Reality, an experience which he describes as "wholly non-rational" and relating to the "wholly other." It may be accompanied by energy or urgency, a "consuming fire" of vitality, passion, and excitement, and "over-brimming" religious emotion. The fascinating aspect of religious experience tends to involve less intensity and be generally positive. Otto describes the state as "wonderfulness and rapture," bliss, transport, glory, and the "beatific experience which is found in the religious feelings of longing, exaltation, serenity and joy."[4]

This "immediate and first hand religious emotion" gives insight into the deeper aspects of the self, but it does not initially include any analysis of that state. Because of this, he differed with Friedrich Schleiermacher's concept of "creature-feeling," which Schleiermacher understood as the major religious emotion, the "feeling of dependence." Such a feeling of dependence would require someone to be dependent upon, such as a God. Otto felt that a person determining the nature and origins of the experience involved rational deduction, and thus included a secondary event.[5]

Otto added an evolutionary theory to this set of descriptions. He wrote that the direct experience of the numinous was a "primitive" form of experience which evolves into religious institution with the addition of ethical systems and philosophical thought. Thus, we have the "early stirrings of numinous consciousness" which move on to higher monotheistic religion through the processes of rationalization and moralization. As the evolutionary theories of religion came to be discredited later in the twentieth century, Otto's work ceased to be cited in books dealing with ecstatic states. However, his ideas on the numinous were influential for many writers. He was willing to examine non-Western religious experiences at a time when this was rare, and to include them as legitimate forms of religious ecstasy, not simply equating them with the Devil's delusions as did many writers on religion of his time.

William James (1842–1910) was a physician, a psychologist and a philosopher. He published his book *The Varieties of Religious Experience* in 1902, and his ideas on ecstatic religious experience have influenced thinking in the field for over a century.[6] Throughout the book, James speaks in favor of inner experience, especially the sort of ecstatic religious experience that is intense and novel and differs from the "imitative" sort of religion which he finds prevalent. As he states, we should not focus on the ordinary believer, whose religion has been made for him by others, but rather on religious "geniuses" whose experiences influence and determine others. In his well-known phrasing:

> It would profit us little to study this second-hand religious life. We must make search rather for the original experiences which were the pattern-setters to all this mass of suggested feeling and imitated conduct. These experiences we can only find in individuals for whom religion exists not as a dull habit, but as an acute fever rather.[7]

James studies such individuals throughout the book, but primarily in his lectures 16 and 17 on mysticism. While he rarely uses the term "religious ecstasy," he often refers to "mysticism" or "mystical states." Realizing that there were problems in definition, he gives four qualities or "marks" by which one may judge ecstatic or mystical states (which he noted were primarily to avoid those understandings which dealt with telepathy or returning spirits). These qualities which define a mystical experience are ineffability (it requires direct experience, and defies expression); noetic quality (it gives truth); transiency (it is brief); and passivity (the person feels as if grasped by a greater force).[8]

James was also interested in both triggers of these states and types of mystical states. He supports the use of "intoxicants and anesthetics" as ways of generating mystical states of consciousness, especially the use of nitrous oxide. While the drug states can give insight, and "depth beyond depth of truth seems revealed to the inhaler," it is not a lasting truth.[9] What is most significant about these states is that they give us knowledge that is different from our normal waking consciousness, which "is but one special type of consciousness, whilst all about it, parted from it by the filmiest of screens, there lie potential forms of consciousness entirely different."[10]

James notes that we may go through life without suspecting the existence of these alternate states of consciousness, but they are nevertheless there, awaiting the requisite stimulus. They are difficult to study, as we have no maps, but we need to include them in any full understanding of the universe. As he phrases it, they "forbid a premature closing of our accounts with reality."[11] While such states can include "ecstasy of vision" and the "immediate presence of God," what seems to be most important to James is what they say about human perception and cognition. Mystical insight "resembles the knowledge given to us in sensations more than that given by conceptual thought."[12] This is knowledge by immediate feeling and intuition. James notes that its cognitive aspects may range from the vision of the future and the reading of hearts to the supersensible forms of theological and metaphysical revelation, and as in St. Teresa's life, the highest states of ecstasy give us knowledge of God.[13]

While the medical mind of his time generally looked at "ecstasies" and saw hypnoid states, superstition, and degeneration, James found that there are many positive aspects to such states. These include energy, optimism, a sense of depth and meaning, and a feeling of unity with the Absolute.[14] Such a positive outlook would be of great personal importance to James, who had a tendency toward depression. He also notes that such experiences relativize non-mystical states as "the sole and ultimate dictators of what we may believe."[15]

James recognized that religions have been subject to "historic aberration" and fanaticism, but he did not equate such events with mystical ecstasy. He felt that the fruits of religion were "liable to corruption by excess," and religion could become pathological when it limits intellect and comes to emphasize fanatical loyalty, jealousy over the deity's honor, and a preoccupation with imagined slights upon the deity. When this is combined with imperialistic churches, and intolerance and persecution, we have a dominance of feeble intellects and aggressive personalities as a part of religious fanaticism. This was a darker form of ecstatic state, which James called the "theopathic" condition, one which rejects knowledge and is imbalanced.[16]

James was interested in what ecstatic states could say about the human mind, and what sorts of alternative types of knowledge were possible. His ideas came to influence the academic study of religious experience, and also Alcoholics Anonymous and other of the Anonymous movements (through the idea of a "higher power" which may be experienced and construed in a variety of ways). Mystical ecstasy could be understood as a way to study extremes of both knowledge and emotion, for the extremes of experience give us insight into the means or more common sorts of functions.

Another approach to the study of religious ecstatic states can be found in the work of Carl Gustav Jung (1875–1961). He was a psychiatrist and initially a student of Sigmund Freud. He agreed with Freud's theory of the personal unconscious containing memories and traumas, but Jung also came to believe in the existence of a deeper collective unconscious containing universal archetypes.

Jung's interpretation of spiritual development was based on an expanded psychoanalytic model of the mind. In times of stress, or of spiritual growth, the self or psyche moves inwards through the personal unconscious with its memories and complexes, into the mythic themes and symbols of the collective unconscious. When the psyche has been fragmented or damaged, this descent allows it to be reorganized and healed. He called this entrance into the depths of the mind Individuation, which involves a "drawing inwards" of libido or psychic energy. The person experiences this as a "magnetic pull" which draws the psyche back to its origins in the deeper layers of the mind. There several things could occur. The psyche may be overwhelmed and thrown into crisis. However, if an archetypal symbol (especially the archetype of the Self) emerges, the old identity is let go, and a new one is born. As Jung phrases it, the psyche "plunges into the fountain of youth," and is remade. This is ecstasy for Jung, when the older structure of the self is reorganized, and it is then exhilarated by the influx of powerful psychic energies. The ego is not merged with the Self as the archetypal God symbol, or there would be the danger of inflation (with the ego identified with the God, and unable to separate the two). The individual self must become a separate center of meaning and inspiration.

Jung used the term "psyche" to include the totality of all psychic processes. Jung's ideas on the collective unconscious changed over the course of his writings, and he has described it according to biological, genetic, and psychological models. One important aspect of experience of the collective unconscious is the process of visionary transformation. As he notes, visions are important, for the psyche consists primarily of images, and indeed "every psychic process is an image."[17] The ego becomes aware of depth through images, seen as it is pulled inwards. If the Self archetype is directly encountered, the old ego dies, and the psyche is transformed into a new state, with the "power of the gods." It brings the vision of eternal life, and the person is no longer paralyzed with depression and fear, and is able to go forward into the unknown. In mythic terms, this is the sacrifice of the old ego, sometimes symbolized by the battle with the dragon in the cave, and the descent into hell. The psyche can then encounter symbols of the Self archetype: the mandala, the jewel, the child, the hermaphrodite. It must be more than simply a personal breakthrough, for "authentic individuation must ultimately extend to the universal."[18]

The Self is experienced as the source of life for the ego, as an authoritative presence and the axis around which it revolves. Though the Self becomes the center of the personality, it is not identical to the ego. As Jung notes, "The Self has as much to do with the ego as the sun with the earth. They are not interchangeable."[19] The Self appears as a God-figure, and the encounter with it is often understood as a mystical experience. This occurs temporarily in the vision of wholeness, when the consciousness is fused with the archetype of the Self and its structures are reconfigured to fit the archetypal pattern. As Jung notes, "In this way conscious and unconscious are interfused, and a decided change of consciousness is brought about. This is why I call this a process of transformation."[20] He compares this process to Christ's descent into hell:

> The three-days descent into hell during death describes the sinking of the vanished value into the unconscious, where, by conquering the power of darkness, it establishes a new order, and then rises up to heaven again, that is, attains supreme clarity of consciousness.[21]

While this ecstatic fusion is positive in small doses, bringing a "regressive restoration" which helps the psyche, Jung is unwilling to defend it for long periods of time. He finds mysticism of this type too dangerous, especially for the Western mind, and he encourages the return to ego-consciousness as soon as possible. Both the ego and the Self must be healthy for the person to be able to function in the world.

Jung felt that all neuroses expressed a disturbance of the religious function of the psyche, and the acting-out of a loss of psychic balance and sufferings compel the person to come to terms with the foundations of being and the world.[22] The goal of ecstatic visionary experience is a return to the world as a healthy and psychologically balanced person. Without such experience, psychic energy can assume negative forms, leading to at the social level to genocide, holocaust, war, and other disasters. At the personal level, the lack of such experience brings a life that is deadened, without purpose, causing the person to struggle with despair, suicidal impulses, and addiction. Thus, ecstatic and mystical experiences held an important place for Jung, as major means of psychological healing. However, we should note that he avoided traditionally religious interpretations of such events for a specific reason: he had not been trained as a theologian and wished to stay within what he considered to be the areas of his training.

As for religious ecstasy as ontology, we can look at the writers Paul Tillich and Mircea Eliade. Ontology is the philosophical study of the nature of being, becoming, existence, or reality, as well as how these are organized, a study that goes back to Plato and Parmenides in ancient Greece. Today, it is sometimes conflated with Theology. However, ontology does not require a specific *theos* or God, it simply analyzes the way the world can be understood to exist. Theology is the understanding of a religion by its followers, and ontology does not require a faith commitment. For both Tillich and Eliade, religious ecstasy reveals a hidden dimension of awareness, which can be understood as a God, but need not be equated with one.

For Paul Tillich (1886–1965), ecstatic states reveal the "depth dimension" of existence. This dimension is not limited to institutional religion, as it can be found in any field as depth of meaning or "ultimate concern." It is by intense subjective experience that this dimension of life is encountered. Tillich speaks of the vertical dimension of life, revealed primarily through symbols. These may start out as ordinary or dead symbols which give only information, but they may be empowered or enlivened by intense experience. This allows the symbol to become transparent to the light within it, to open up and reveal meaning to the person experiencing it. A symbol in the horizontal dimension reveals only information, and a symbol in the vertical dimension reveals deeper truth.

While Tillich was both a philosopher and a theologian, he often wrote from one perspective or the other. As Tillich notes in "Religion as a Dimension in Man's Spiritual Life," the question of whether God does or does not exist is not a relevant religious question. "A God about whose existence or non-existence you argue is a thing beside others, within the universe of existing things."[23] He finds this to be a mistake made by both scientists and theologians, who argue with each other over whether revelations are authoritative. Tillich finds religion to be an aspect of the human spirit, not a belief in a God or a debate over such a belief. In this way, religion surpasses Theology, which requires a deity.

For Tillich, religion is a "point of view," which allows a vision of the depth of human spiritual life. It is associated with morality, with cognition, with mythological imagination and with aesthetics, but it is not limited to any of them. He understands the metaphor of depth to refer to that which is "ultimate, infinite, unconditional" in life.[24] It is shown in different ways in different aspects of life. In ethics, it is "the unconditional seriousness of the moral demand"; in knowledge, it is "passionate longing for ultimate reality"; and in aesthetics, it is "the infinite desire to express ultimate meaning." He finds that all of these are ways of encountering spiritual depth.[25] Ecstatic experience reveals these depths, overcoming alienation and estrangement from the world and from other people. Religious ecstasy gives us "the experience of the Holy, of something which is untouchable, awe-inspiring, an ultimate meaning, the source of ultimate courage."[26]

However, religions forget their own depths, and this is their problem. They make myths and rituals and laws into ultimates, and persecute those who do not follow them. They forget their own origins, and this alienation causes much of the hostility of the secular world towards religious groups. Thus, the lack of religious ecstasy has had tragic consequences for both religion and the secular world. It is ecstatic experience which gives depth and purpose to life, allowing people to overcome despair and alienation. Without it, life loses its meaning, and the world becomes flat and finite.

Another writer who explored the ontological aspects of ecstasy was Mircea Eliade (1907–1986). Eliade described the ways that ecstatic states as both hierophany and gnosis give insight into the deeper dimensions of life. He contrasts sacred and profane realms of existence, and his concern is the way that the universe exists. In his book *The Sacred and the Profane*, he describes the sacred as an "existential dimension" which is different from the desacralized world, which allows the person to "participate in reality."[27] The person encounters this aspect of the world through breaks or interruptions in profane space, which allow the sacred space to be perceived. This encounter with sacred space comes through an ecstatic experience, a *hierophany*, which reveals the limitations of ordinary profane space. He notes, "Every sacred space implies a hierophany, an irruption of the sacred that results in detaching a territory from the surrounding cosmic milieu and making it qualitatively different."[28]

This is not a study of Theology, of a God and his behavior, but rather a study of how the world is constructed. Like Tillich, he understands the sacred in almost geographical terms, as a dimension or a realm. Revelation does not require a cosmic person, but rather space: "The sacred reveals absolute reality and at the same time makes orientation possible; hence it founds the world in the sense that it fixes the limits and establishes the order of the world."[29] The state of religious ecstasy is the linking of these worlds together, an insight into the structure of creation. While sacred space may be understood and experienced in a personal fashion, through a theophany or a revelation of a God, his concern is the nature of the space, rather than the actions of a particular God.

Eliade did not focus upon the emotional aspects of ecstasy, as Otto did, but rather looked at its effects on perception and sense of self. Ecstasy brought *gnosis*, spiritual knowledge, based on the experience of travel between sacred and profane worlds. Eliade also associated ecstasy with shamanism, which he understood as a worldwide tradition in which religious specialists gained the ability to make spiritual journeys into other worlds. The shaman could ascend into the heavens or descend into the underworld, and his or her abilities came from strenuous efforts at training and from initiation. In his book *Shamanism: Archaic Techniques of Ecstasy*, Eliade describes how the shaman gains control of ecstatic states and can use them to accomplish his goals. He finds religious ecstasy is central to shamanism:

> The essential and defining element of shamanism is ecstasy- the shaman is a specialist in the sacred, able to abandon his body and undertake cosmic journeys "in the spirit" (in trance).[30]

In his book on yoga, Eliade differentiates between ecstasis and enstasis. While yogic literature often includes such ecstatic elements as magical flight and liberation of the soul, as well as some "momentary, flooding ecstatic states with colors," its goal of *samadhi* is really "stasis" rather than flight, a still and unchanging elevation of the mind and tranquillity of the heart.[31] Mental function ceases, yet there is full comprehension of being, accompanied by the end of suffering. While ecstasy is active and transformative, enstasy is still.

For Eliade, ecstasy is a sort of skill, a technique which shamans and others learn to control, which gives a special kind of knowledge, and a new identity when the old self dies and a new self is born. It is a legitimate means of learning, found in many cultures and religions around the world. As in the case of Tillich, it gives insight into the depth dimension of human life, which may be understood through a variety of symbols and myths.

There are also modern writers who have been interested in ecstatic states in a more nuanced way. Some books examine the historical impact of ecstatic states, such as Ann Taves' *Fits, Trances and Visions: Experiencing Religion and Explaining Experience from Wesley to James*; Leigh Eric Schmidt's *Restless Souls: The Making of American Spirituality from Emerson to Oprah*, and Eugene Taylor's *Shadow Culture: Psychology and Spirituality in America*. All of these writers give useful perspectives on the history and development of ideas on religious ecstasy, and they describe its relevance for both religion and culture.

But for modern, more popular justifications of the study of ecstasy, we have two areas that have been emphasized in recent years, the pragmatic and the theoretical. From the pragmatic approach, the relevant question is whether ecstatic experience can affect people for the better in the physical world, improve their lives, and add to health and happiness. We see the pragmatic approach used in healing addiction and trauma, following the concept of ecstasy as wellness and transformational healing that was popularized by Abraham Maslow as a goal of "peak experiences."

From the theoretical approach, the most important question is whether ecstatic experiences can give new forms of knowledge, solutions to problems and new creative insights into the world. We see this area explored in sociology, and transpersonal theory, and in modern esotericism. Andrew Greeley and Ken Wilber have written about ecstatic knowledge, and we shall count them as popularizers of this approach. The important question here is, can we learn anything valuable from religious ecstasy? Modern research has examined both of these areas, and these are both directions for the study of religious ecstasy which can incorporate interdisciplinary data.

Ecstasy Deprivation and the Pragmatic Argument

The pragmatic argument about religious ecstasy states that ecstasy can have social impact, and it can accomplish useful things for both individuals and societies. It further argues that modern Western culture lacks tolerance for ecstasy, and that it suffers as a result. We can see this in such groups as Alcoholics Anonymous, and in humanistic and transpersonal psychologies.

The claim that the deprivation of religious and ecstatic experience drives many social maladies has been influential in the popular culture of the twentieth century. Perhaps its most famous advocate was Bill Wilson, whose ecstatic "nighttime spiritual epiphany" became the original inspiration for the group Alcoholics Anonymous. Wilson was inspired to found AA by his own mystical experience. During his fourth admittance to Towns Hospital in 1934, he recalls his "white light" experience: "Suddenly the room lit up with a great white light. I was caught up into an ecstasy which there are no words to describe. It seemed to me, in the mind's eye, that I was on a mountain and that a wind not of air but of spirit was blowing. And then it burst upon me that I was a free man."[32]

Wilson believed alcoholism to be a symptom of a deep unfulfilled need, which could be satisfied by anything from the ecstatic experience that personally changed him to community among lonely souls. Both the wide variety of Anonymous and Twelve Step groups and many psychological writers claim that the cure for addiction can come from any religion or contemplative path, if it genuinely brings about an ecstatic state. As an example, the Jungian therapist Marion Woodman notes,

> Many people in our society are being driven to addictions because there is no collective container for their natural spiritual needs. Their natural propensity for transcendent experience, for ritual, for connection to some energy greater than their own, is being distorted into addictive behavior.[33]

Christina Grof describes her motivation to use alcohol as a search for the "promise of freedom, connectedness, and love" she had experienced as glimpses throughout her life. She says,

> And I *knew* I found it in the delicious oblivion of alcohol. My boundaries melted, the pain disappeared, and I was, I thought, free. I felt comfortable within my own skin and felt connected with a carefree vitality that told me I could do anything. I was at ease with people in a way that was impossible in my daily life. I felt included, accepted, and cherished – until alcohol turned against me.[34]

The "thirst for wholeness," as Christina Grof entitled her book about addiction, was the underlying motivation behind her desire for alcohol. The title comes from Carl Jung's statement referring to the craving for alcohol as the "spiritual thirst of our being for wholeness, expressed in medieval language: the union with God."[35] Grof has described the goal of her own alcoholism as "the direct encounter with mystical or transpersonal realms."[36]

Alcoholics Anonymous, the first of the Twelve Step groups, followed Carl Jung's phrase, "*spiritus contra spiritum*."[37] As Jung wrote, "alcohol" in Latin is *spiritus*, and people use the same word to refer to the religious experience of the Holy Spirit and to alcohol. The formula therefore translates into the principle of using spirituality against the addiction to alcoholic spirits; substituting God (in whatever form that has meaning for that individual) for alcohol. The researchers McPeake, Kennedy, and Gordon argue that this need is a missing component of most psychologically based addiction treatment programs. They claim that without the opportunity to experience altered and especially ecstatic states in "constructive, ritualized, socially approved ways", individuals find other ways to alter consciousness which may be destructive to themselves, their loved ones, and their communities.[38] The various Twelve Step programs, with their language of hitting bottom and the revelation of dependence on a higher power, have been adapted to many problems. We now also have Gamblers Anonymous, Narcotics Anonymous, Emotions Anonymous, and Overeaters Anonymous, among many others, as well as various survivor, family, and co-dependency groups. These are described in the Twelve Step literature as "spiritual maladies," whose cure is "spiritual awakening," and millions of people have joined these groups. Some understand the higher power as social, the presence of other supportive people, but many understand the higher power as spiritual insight which can be revealed in ecstatic experience, much as Bill Wilson described.

As the analyst Robert A. Johnson writes, the alienation that motivates addiction and other such social pathologies is due to the lack of access to ecstasy in modern society:

> Ecstasy. It was once considered a favor of the gods, a divine gift that could lift mortals out of the ordinary reality and into a higher world. The transformative fire of ecstasy would burn away the barriers between ourselves and our souls, bestowing on us a greater understanding of our relation to ourselves and to the universe. It is the great tragedy of contemporary Western society that we have virtually lost the ability to experience the transformative power of ecstasy and joy. This loss affects every aspect of our lives. We seek ecstasy everywhere, and for a moment we may think we have found it. But, on a very deep level, we remain unfulfilled.[39]

More specifically, that lack of fulfillment motivates addictive behavior:

> Addiction is the negative side of spiritual seeking. We are looking for an exultation of the spirit; but instead of fulfillment we get a short-lived

> physical thrill that can never satisfy the chronic, gnawing emptiness with which we are beset. To fill this emptiness, we need to reconnect with the capacity for ecstasy that lies dormant within us. Our first step must be to try to understand the nature of ecstasy.[40]

Addiction, in its various forms, continues to be a problem into the twenty-first century. As to its link with the lack of ecstatic states, some of the best data might be seen in the popularity of the various Anonymous and Twelve Step programs, where people are able to discuss their intense experiences (from "hitting bottom" to becoming new people), in sympathetic groups.

We also see ecstatic states in their milder forms being encouraged in the growing popularity of Vipassana meditation, especially in its Western renaming as "mindfulness meditation" or "awareness meditation." It is advertised much as Transcendental Meditation was in the 1960s and 1970s, as something to help with stress, high blood pressure, and heart problems, bringing serenity and insight. But such meditation can induce ecstatic states as well, which is part of the way that anxiety is calmed and changed into appreciation. In Vipassana meditation, ecstasy is specifically located within the *jhanas* or deeper states, but even initial states can bring milder ecstasies of joy and appreciation of the beauty of the present moment. The meditation ideally brings accurate understanding of experience and insight into reality. This results in inner peace, altruism, and spontaneous benevolence.

Some people can combine several paths. As Jay Michaels describes meditative ecstasy, "There are, really, only two choices available to someone who wants to suck the marrow out of life: either continually seeking more extreme experiences, or making *every* experience 'extreme.' Some people can apparently do the former, but I find that tiring. Meditation allows me to 'suck the marrow' out of each tree, table, soda, or breath. By eliminating signal noise and stopping thought, the true colors of the phenomenal universe become revealed, in ever-increasing brilliance."[41] He notes the importance of such altered states:

> Many people deeply fear altered states of consciousness, I think because they are overly afraid of their own non-rational minds. Subscribing to a worldview in which 'rational' rules of decency, propriety, etc., govern every aspect of life means relying on our capacities of rational judgment for every important decision. And so, mind-states which relegate such faculties to

> a subordinate or even invisible role is scary. Now, of course, I'm all for rational judgment making most decisions in the world, and certainly all of those which seriously affect other people. But is it a rational judgment to dance? To let go of the self in orgasm? To fall in love? Some of our most transcendent moments come when the rational mind is quieted and something else takes its place. In some aspects of life, being in touch with the nonrational is essential to being human.[42]

While Michaels also has sympathy with the ecstatic states engendered by entheogenic drugs, he has come to prefer the meditative states which come from Vipassana meditation. However, he avoids claims of "spiritual supercession." Rather than avoiding drugs as potentially addictive, as we see with the Anonymous groups, he finds that they can enrich spiritual knowledge if used to deepen meditative practices.

Michaels notes the importance of "being in touch with the nonrational." As the therapist Jalaja Bonheim notes in her book *The Hunger for Ecstasy*, ecstatic states are necessary to a healthy life,

> We all need ecstasy in our lives. That state of rapture, of bliss, of feeling totally in love with life is an essential nutrient without which we cannot thrive. Yet if ecstasy is spiritual food, we are a nation of starving people. Children of a culture that ignores the hunger for ecstasy, we know no other way of life and therefore assume that our ecstasy deprived state is normal.[43]

Her claim is that the deprivation of ecstasy is the basis of many social problems: depression, drug addiction, violence, and broken marriages. This is based on her clinical work, in which clients describe a sense they are missing out on some aspect of life, and try to find it through unhealthy alternatives. "If we want to build a healthy, peaceful society, we must face the fact that we need joy, rapture, ecstasy in our lives as much as we need food."[44]

As journalist and cultural critic Barbara Ehrenreich notes in her book *Dancing in the Streets: A History of Collective Joy*, ecstasy used to be a part of popular culture through music and dance. However, over the past few centuries, there have been condemnations of both ecstatic states and creative expression. European colonizers were appalled by the energy that natives invested in superstitious and ritual activities, and possession dance was a symbol of loss of control. While Calvinism sought to restructure the person into being ascetic, self-denying, and single-minded, dancing violated this self-control and avoidance of sin.[45] Victorian sensibilities avoided any ecstatic dimension to sexuality as too close to animal behavior, and as wild music could also stir savage passions, it too needed to be controlled. By the 1980s, the rebellious music of rock was commercialized and bowdlerized into Muzak; it was made into a commodity and used for marketing everything from cars to financial services. As the music was controlled by the culture, so was the ecstatic state it could engender. Companies appropriated defiance in commercials for newness and rebellion, ecstasy was subverted into becoming a marketing tool, in order to sell products. Video games simulate drug experience, but control it—there is no opportunity for religious ecstasy, only the joy of competition.[46]

Ehrenreich writes about the "disdain" for ecstasy by scholars and intellectuals of the West, and the apparent incompatibility of ecstatic states with modern civilization. She argues that one reason for this is their background of American conservative Protestant values, their focus on "hard work, sobriety, and meager forms of charity," their glorification of the military, and their religious rituals which are "pallid affairs" requiring faith rather than experience. There is a lack of ecstatic rituals which are "expressive of our artistic temperament and spiritual yearnings."[47]

Ehrenreich cites the French theorist Guy Debord, who claims that in the modern West we have spectacle instead of festival. People do not generate their own pleasure; instead, they absorb or consume spectacles of commercial entertainment, consumer culture, and nationalist rituals. Entertainment is passive, walking in malls, sitting in movies, and listening to iPod music. We have anti-depression drugs rather than pro-creativity drugs.[48] Ecstasy is vicarious, watched rather than created. Ehrenreich wishes to redeem ecstasy for the modern world; she writes, "Why not reclaim our distinctively human heritage as creatures who can generate their own ecstatic pleasures out of music, color, feasting and dance?"[49] Her pragmatic argument is that ecstatic states enrich cultures and individual life.

Ecstasy can also be used toward the practical goal of psychological healing. One approach to ecstatic healing has been through the "peak experiences," popularized by Abraham Maslow. Another has been psychedelic therapy, in which individual ecstatic states are induced by the use of psychedelic drugs in order to deal with trauma.

A popularizer of the idea that ecstatic states could help people in responding to life's problems, Abraham Maslow described the concept of "self-actualization," in which a "peak experience" or "core religious experience" could bring meaning and inspiration to an individual or to a culture. Maslow's interpretation of ecstasy as a "peak experience" has become one of the most well-known popular understandings of ecstatic states. The idea of ecstatic peaks as a part of human development or "self-actualization" became so popular that its associated phrase "be all that you can be" was taken over as a slogan for the US Army.[50] Peak experiences were ideal states, transformational states, and the best possible states for people to experience. Maslow wished to focus people on the future and its positive experiences, rather than the neurotic past, and he believed that these experiences could inspire new forms of thinking and feeling. Maslow was president of the American Psychological Association (from 1967 to 1968), he founded the *Journal of Humanistic Psychology* (with Anthony Sutich) and also the American Association of Humanistic Psychology. Maslow's ideas are found today in the field of "positive psychology" and modern research on happiness.

In his book *Religions, Values and Peak Experiences* Maslow discusses the immediate, subjective responses to such experiences. It was Maslow's claim that the "core experience" of every known high religion is found in the personal illumination or ecstatic revelation of a prophet or seer.[51] Ecstasy was the universal nucleus of religion, the essence of which is the "peak experience" or "transcendent experience." He believed that all revealed religions began in ecstatic revelation, and that all of their "peak experiences" were the same in essence, though they differed in their cultural expression and symbolism, what Maslow calls "localisms" or accidents of particular languages and philosophies. Indeed, each "peaker" develops his own religion out of his or her private ecstatic revelations, and Maslow notes that the "paraphernalia of organized religion"—buildings, personnel, and dogma—are of secondary and doubtful value. Religion could be developed independently, and not simply accepted as a historical tradition. People who had peak experiences had a different sort of mind as well, and he contrasts their "positive psychology" with the "psychopathology of the average," thus reversing the dominant view in psychology.[52]

Maslow describes peak experiences as "rare, exciting, oceanic, deeply moving, exhilarating, elevating experiences that generate an advanced form of perceiving reality, and are even mystic and magical in their effect upon the experimenter." Maslow believed the origin, core, and essence of every known "high religion" was "the private, lonely, personal illumination, revelation, or ecstasy of some acutely sensitive prophet or seer."[53] In this way, he echoes the earlier views of William James.

People who are able to have ecstatic or peak experiences are understood by Maslow to have a different way of thinking, which he calls "B-cognition," or thought based on awareness of ultimate Being. This allows the person to see the world "under the aspect of eternity," and he or she experiences unitive consciousness. Such cognition involves great concentration, and a non-evaluating and non-judging ecstatic appreciation, the way that a loving mother might look at her newborn infant. It is a perception of the realm of Being, in which the universe is viewed as an integrated whole. The world becomes beautiful and blissful, and person experiences such emotions as pity, charity, kindness, wonder, and reverence. This attitude is detached, objective, and unselfish, less petty and critical than ordinary consciousness. It makes life worthwhile, a thing of intrinsic value, and allows for the acceptance of death without fear and transcendence of conflicts. Even evil is relativized as a part of the world order.[54] All things are made new, and attention is widened to perceive the whole cosmos as a unity.

But when ecstatic states are limited by traditional religious contexts, they become empty and valueless. Many people lose or forget the subjectively religious experience, and end up under the sway of institutional religion. He defines such religion as "a set of habits, behaviors, dogmas, forms, which at the extreme becomes entirely legalistic and bureaucratic, conventional, empty, and in the truest meaning of the word, anti-religious."[55] Mystical and ecstatic experience, the "illumination of the great awakening," is later forgotten and transformed, with organized religion as the enemy of the religious experience because it claims power over it.

As a psychologist, Maslow was interested in the psychological effects of such states. He argued that ecstasy works against alienation, emptiness, hopelessness, and depression, and brings good psychological health. It brings "B-love," or love which is based on awareness of ultimate Being. Such love is unselfish and understanding, admiring the intrinsic nature of others, and the "peaker" appreciates other persons as ends in themselves. There is a sense of mystery, gratitude, wonder, joy, rapture, bliss, and ecstasy. Such emotion rests upon "B-values," values which are based on the experience of ultimate Being. These include truth, goodness, beauty, wholeness, aliveness, uniqueness, justice, order, playfulness, autonomy, and completeness.[56] He believed that human life needs more of these values. The opposite approach to the world, according to Maslow, is working on the basis of "D-cognition," or thinking based on an assumption of deficiency. Those who view the world as deficient and imperfect are full of anxiety, worries, fears, and pains, and they understand the world to be profane, evil, cold, and temporary.[57] While this approach may be accurate for specific persons in specific situations, the more important question is the nature of human life itself, and whether it is basically good or basically evil.

Maslow believed that the forces in societies which work against ecstatic emotion and perception are harmful, they diminish, pathologize, cause regression, and reduce all thinking to the concrete. This would include both the strict materialist perspective and also the perspective of many forms of institutional religion. For Maslow, such people are "not fully functioning," not sufficiently integrated, and living in a desacralized world.[58] As a result, their lives are less happy and fulfilled. Peak experiences, on the other hand, are "rare, exciting, oceanic, deeply moving, exhilarating, elevating experiences that generate an advanced form of perceiving reality, and are even mystic and magical in their effect upon the experimenter."[59]

Toward the end of his life, Maslow determined that humanistic psychology could explore many forms of human potential, but it was not sufficient for understanding states of religious ecstasy. Another form of psychology was needed, which could explore the interface between psychology and spirituality, and accept spirituality as a legitimate area of academic study. He argued that there was another, higher level of development, what he called *transcendence*, associated with altruism, spiritual awakening, liberation from egocentricity, and ultimately the unity of all being. Such transcendence "refers to the very highest and most inclusive or holistic levels of human consciousness, behaving and relating, as ends rather than means, to oneself, to significant others, to human beings in general, to other species, to nature, and to the cosmos."[60] This exploration of transcendence became a basis for the field of transpersonal psychology.

While Maslow's work has largely disappeared from many academic psychology programs, it can still be found in positive psychology (it has been linked with Mihaly Csikszentmihalyi's concept of "flow"), and it has been influential in New Age thinking. However, modern psychology has largely dropped the religious imagery and metaphors. The Fall 2010 issue of *Psychology Today* had a cover story on "Big Moments: Chasing Peak Experiences." It described modern researchers who are interested in the "sudden feelings of intense well-being that fill us with wonder and awe."[61] They do research following a secular model, in which the ultimate state is not described using religious symbolism. It is rather an ultimate extreme of happiness, transcending ordinary awareness, "times when we feel extraordinary and our experiences are recorded in Technicolor." These sorts of experiences are memorable and life-changing. The *Psychology Today* article refers to peak experiences as "epiphanies," a term of religious origin, which is here defined as an "aha! moment." Such a moment can reorder priorities, reveal the person's authentic self, and inspire behavior that fits the newly acquired values. Researchers have simulated epiphanies in laboratory settings by using neuroimaging techniques to watch subjects solve difficult problems. At that moment, visual activity is reduced, and attention is focused. William Miller at the University of New Mexico calls it a "quantum change- when a sudden realization leads to an immediate behavioral reorganization."[62] This is different from ordinary behavioral change, which tends to be gradual. However, this is understood as a psychological rather than a spiritual event.

As Jeffrey Kottler at California State notes, "A lot of peak experiences involve a feeling of spiritual transcendence, not necessarily in a religious way, but where you feel transported beyond yourself as a human being."[63] This is perhaps as close to the language of religious ecstasy as modern psychological researchers are willing to go. The article gives suggestions for inducing peak experiences, including volunteering, altruistic acts, and taking on challenges. Such induction of ecstatic states allows for the study of variables, such as time, place, local influences, and imagery.

In this way, we may note that it is particularly different from the constructivist view of ecstatic experience, which does not allow for variables. As each culturally defined ecstatic state would be totally different from those described in other cultures and teachings, we cannot have any scientific study involving comparisons. Social scientific research is normally based on the analysis and comparison of events, and is not prevented by differing belief system, histories, and culture. This unwillingness to have any continuum for comparison is a limitation of the modern field of Religious Studies.

The area of transpersonal psychology, which Maslow came to value toward the end of his life, came into popularity as a way to understand the effects of both religious meditation and psychedelic drugs. It came to be treated with skepticism and even hostility after LSD became illegal (especially after the drug became associated with the counterculture and protests against the Vietnam war). It is currently returning slowly to legitimacy, going back to the earliest stage of research, before psychedelics became commercialized and controversial.

Perhaps the most important voice in transpersonal psychology today is Stanislav Grof, a psychiatrist with over sixty years of experience in research of non-ordinary states of consciousness and one of the founders of transpersonal psychology. He was born in Prague, Czechoslovakia, where he was trained as a physician and he underwent training as a Freudian psychoanalyst. Grof explored the therapeutic potential of psychedelic substances at the Psychiatric Research Institute in Prague, and since that program, he has observed thousands of patients who have taken LSD. In 1967, he became Clinical and Research Fellow at Johns Hopkins University, and he later became Chief of Psychiatric Research at the Maryland Psychiatric Research Center. He developed a set of maps of the deeper levels of the human psyche based on the experiences of his subjects on LSD, and later developed a technique for exploring deeper states of mind without the use of psychedelic drugs, called Holotropic Breathwork.

Grof initially defined ecstatic states as "non-ordinary states of consciousness," and later as "holotropic," thus "oriented toward wholeness" or "moving toward wholeness" (from the Greek *holos*, meaning whole, and *trepein* as moving toward or in the direction of something). He notes that "Holotropic states of consciousness have the potential to help us recognize that we are not 'skin-encapsulated egos'—as British philosopher and writer Alan Watts called it—and that, in the last analysis, we are commensurate with the cosmic creative principle itself."[64] While earlier such non-ordinary states were called altered states of consciousness, he dislikes the term (associating it with veterinary medicine and the castration of dogs and cats).

For Grof, religious and spiritual forms of ecstasy can be induced by the use of psychedelic plants, such as ayahuasca, peyote, and psilocybin mushrooms. Research on induced states of religious ecstasy involved contact with various North American, Mexican, South American, and African shamans and healers. He also worked with representatives of various spiritual disciplines, including Vipassana, Zen, Vajrayana Buddhism, Siddha Yoga, Tantra, and the Christian Benedictine order. Non-ordinary states can become states of religious ecstasy when they use the imagery and practices of religious disciplines. He accepts that many ecstatic states will remain anomalous phenomena, as we do not yet have ways of understanding them. But an important aspect of religions is the way that they can guide and influence ecstatic states.

Such ecstatic states allow people to explore the deeper levels of the mind, which Grof calls the transpersonal levels. These include a wide array of experiences in which consciousness transcends the boundaries of the body/ego and the usual limitations of linear time and three-dimensional space. Such ecstatic experiences result in identification with other people, groups of people, other life forms, and even elements of the inorganic world. In his book *Realms of the Human Unconscious*, Grof distinguishes a variety of types of ecstasy, such as oceanic ecstasy and volcanic ecstasy.[65] The person may experience a melting, liquid sort of ecstasy with radiance and joy, or a violent and sudden ecstatic experience comparable to a raging storm, a hurricane or an earthquake or a consuming fire. He also discusses ecstatic states by St. Ignatius of Loyola, Sri Ramana Maharshi, St Francis of Assisi, Paramahamsa Yogananda, and others.[66] In this case, the focus is on the religious experience itself, and not on its social impact.

Grof started out as a medical doctor and Freudian psychiatrist, but found that none of the models he had been taught were sufficient for the kinds of experiences he had been observing in his patients. It was necessary for him to incorporate ideas from many world religions to make sense of what his patients were experiencing. For Grof, the exploration and categorization of the transpersonal levels of the mind was an important way to understand states of religious ecstasy. He understood LSD to be an entheogen (from "en-theos," the God within), and he was interested in how mystical states could occur, and their potential for healing trauma.

Grof worked with over a thousand cases of people who had taken LSD during a decade of psychiatric research. He began with a Freudian perspective and gave LSD to clients with mental disorders. He also gave it to mental health professionals, artists, scientists, and philosophers, who were curious about its effects. He found that the Freudian paradigm worked well with low doses of LSD, as clients relived scenes from infancy and childhood. However, with higher doses and repeated sessions, clients moved past the Freudian model into spiritual experiences. Some of these resembled descriptions in sacred texts of world religions with which Grof was familiar.[67] These descriptions by his clients motivated him to develop a new model of the mind. Grof came to perceive the current scientific models of the psyche as "conceptual straitjackets," which confused theory with reality in determining human limitations. He felt that a broader understanding of the psyche was necessary.

In his book *Realms of the Human Unconscious*, he created a "cartography of inner space" which was an alternative mapping of the psyche. It was more inclusive than those accepted by the medical establishment, and included phenomenological descriptions of various levels and types of experiences which appeared in the psychedelic sessions. He noted that in the initial stages, there were visual changes: afterimages, perception of lights, bright colors, geometric forms, and abstract patterns. Sometimes there were exotic environments.[68] After this, psychodynamic imagery would appear. There would be repressed memories, emotional problems, and unresolved conflicts, the traumas of the personal dimensions of the unconscious mind. Clients would reenact events from the past, often following Freudian psychosexual dynamics. Grof noted the importance of a particular set of complexes, which he called COEX systems (or systems of condensed experience). These would follow a particular theme, and the complex would link together all intense or traumatic experiences of a specific type, such as humiliation, or suffocation, or terror. The most important of the systems was the birth trauma, which was re-experienced by many of his clients. He found that there were four stages of the birth trauma (which he called basic perinatal matrices) through which his clients would proceed.

He described ecstatic experiences primarily in the transpersonal realms, which the client would experience after passing through these areas of memory and trauma. He classified them into two types: Experiential Extension within the Framework of "Objective Reality" and Experiential Extension beyond the Framework of "Objective Reality." In the first or "within objective" category, he included two types of experiences: Temporal and Spatial. Under Temporal Expansion, he included Embryonal and Fetal Experiences, Ancestral Experiences, Collective and Racial Experiences, Evolutionary Experiences, Past-Incarnation Experiences, Precognition, Clairvoyance, Clairaudience, and Time Travels. Under Spatial Expansion, he included Ego Transcendence, Identification with Other Persons, Group Consciousness, Animal and Plant Identification, Oneness with Life and Creation, Consciousness of Inorganic Matter, Planetary Consciousness, Extra-planetary Consciousness, Out of Body Experiences, Space Travels, and Telepathy. There was also a smaller category of Spatial Constriction of Consciousness associated with awareness of the physical body, under which he included Organ, Tissue, and Cellular forms of Consciousness.

In the second or "beyond objective" category, he included Spiritualistic and Mediumistic Experiences, such as encounters with Spiritual Entities, Experiences of Other Universes, Archetypal Experiences and Complex Mythological Sequences, Encounters with Deities, Understanding of Universal Symbols, Kundalini Experiences, Consciousness of the Universal Mind, and the Supracosmic and Metacosmic Void.[69] He gives many examples of his patients who describe these states. Eventually, he came to realize that many of these states were described in the myths and folklore of world religions.

Grof spoke in greater detail of his clients' memories of the pre-birth womb state as blissful, peaceful, and unified, a sort of "oceanic ecstasy." He contrasted this with the "volcanic ecstasy" associated with memories of the crushing passage through the birth canal, which involved pain and pleasure, love and aggression, images of death and birth, and a Dionysian style of "wild, ecstatic rapture." These forms of ecstatic experience are thus associated with memories of birth: one type quiet and still, the other active and explosive.[70] Grof's theory is that the psyche has no fixed and absolute boundaries—that it can transcend the limits of space and time under the right conditions.[71] Consciousness is mediated by the brain rather than originating from it, and drugs alter that process of mediation. He argues that our model of the mind needs to be large enough to incorporate non-ordinary experience—ecstatic and mystical states, shamanic experiences, and the sort of states described in Eastern religion. Spirituality is more than pathology, more than superstition and primitive magical thinking. It needs to be analyzed and understood in its own right as *sui generis* or not reducible to psychological or neurological categories and explanations.

R. E. L. Masters and Jean Houston were also researchers in transpersonal psychology. The research of Masters and Houston was based on first-hand observation of over 200 drug sessions, and also over 200 volunteers and psychotherapy patients who had reported on their experiences after taking either LSD or peyote on their own.[72] They included many first-person accounts, from educators, clergymen, attorneys, and other professionals. These cases were described in their book, *The Varieties of Psychedelic Experience.*

Masters and Houston divided the experiences of their clients into four levels of experience. The first is the sensory level, which usually occurs at the beginning of the drug experience. In this state, there is altered awareness of the body and body image, spatial distortion, and perceptual change. The major focus of attention is the sensory world. The second is the recollective-analytic level. At this stage, the person becomes introspective, examining his or her personal problems and life goals. Sometimes the client may relive past events, which are often traumatic. At this stage, it is important for the person to have a guide, to deal with the problems arising from the personal unconscious. Many therapeutic insights may come from his stage. The third is the symbolic level, in many ways comparable to Jung's collective unconscious. Clients would encounter many varieties of symbolism—mythical, legendary, historical, ritual, and archetypal. The person may act out myths and legends, undergo initiation or other religious rituals, or witness historical and evolutionary events unfolding. Cross-cultural themes appear vividly in stories and events, which may be observed from a distance or personally experienced. The fourth is the integral level, and this is very rarely reached. Masters and Houston estimated that only about 5% of the people who took LSD or peyote experienced this level. However, it is immensely important to those who do, for it is a state of ecstatic revelation, illumination, Enlightenment, or direct experience of the divine. It was described by all clients in religious terms, as God, the Ground of Being, the Void, Brahman, or a creative intelligence that pervades the universe. For them, it brings absolute certainty, which is intensely emotional, and results in long-term positive integration of the personality. This state tends to be reached by people with backgrounds in prayer, meditation, or other contemplative practice, mature personalities who were well-prepared for intense experience.[73]

As an example of an ecstatic state which occurred under drug influence, this is the report of a fifty-two-year-old engineer:

> Although consciousness of self seemed extinguished, I knew that the boundaries of my being now had been dissolved and that all other boundaries also were dissolved. All, including what had been myself, was an ever more rapid molecular whirling that then became something else, a pure and seething energy that was the whole of Being. This energy, neither hot nor cold, was experienced as a white and radiant fire. There seemed no direction to this whirling, only an acceleration of speed, yet one knew

> that along this dynamic continuum the flux of Being streamed inexorably, unswervingly toward the One.
>
> At what I can only call the 'core' of this flux was God, and I cannot explain how it was that I, who seemed to have no identity at all, yet experienced myself as *filled with God*, and then as (whatever this may mean) *passing through God* and into a Oneness wherein it seemed God, Being, and a mysterious and unnameable One constituted together what I can only designate the ALL. What 'I' experienced in this ALL so far transcends my powers of description that to speak, as I must, of an ineffably rapturous sweetness is an approximation not less feeble than if I were to describe a candle and so hope to capture with my words all of the blazing glory of the sun.[74]

The Masters and Houston levels are a mapping of the mind, including both its personal and collective dimensions. It was their intention to have such models be useful to people who were exploring these realms, and dealing with past problems and traumas.

The transpersonal approach has been used more recently by psychiatrists such as Michael Mithoefer, in his clinical trials of MDMA as a way to deal with post-traumatic stress disorder, and Roland Griffiths, whose experiments at Johns Hopkins focused on the use of psilocybin to induce ecstatic and mystical states. They are some of the few people today allowed to do psychedelic research by the United States government.

Michael Mithoefer is currently doing research on the drug MDMA (the main active ingredient in the street drug commonly known as Ecstasy) as part of therapy for post-traumatic stress disorder (PTSD). The major treatment now used for PTSD is exposure therapy, when the patient re-experiences traumatic events. Another therapy that can sometimes be helpful for trauma is EMDR, or Eye Movement Desensitization and Reprocessing Therapy.[75] However, revisiting traumatic experiences in therapy may be limited by the patient's inability to tolerate those memories. Recall may cause emotional numbing, which prevents memory and extinction of the fear. A drug that could widen the range between the extremes of fear and numbing improves the efficacy of psychotherapy. In the positive experiences of love and forgiveness associated with the drug, MDMA generally gives four or five hours in which the patient is able to process fear, grief, or rage associated with the trauma without being overwhelmed by negative emotions. It is drug-assisted psychotherapy, in which the drug is not taken alone but while the patient is under the therapist's supervision.

In his high-profile 2011 research, Mithoefer had double-blind studies, which randomized 23 treatment-resistant patients with crime-related PTSD into two groups, which received introductory sessions with a psychiatrist and a psychiatric nurse. Then, all patients went through an all-day experimental psychotherapy session after administration of either MDMA or a placebo.[76] A PTSD Scale for measuring traumatic stress was given at baseline, 4 days after each experimental session, and 2 months after the second session. The rate of positive clinical response was 83% in the active treatment group versus 25% in the placebo group. There were no drug-related serious adverse events or adverse neurocognitive effects. A success rate of over 80% in improvement of symptoms is a very good clinical response. The drug Ecstasy created a state in which the memory of the original trauma can be tolerated, which allows the person to resist the traumatization due to painful memories. Such states may or may not involve religious imagery, but they have the positive emotions associated with ecstatic states. These kinds of results far exceed the success rates for other common forms of PTSD therapy.

Phase III trials for 2018 have been approved for his study, which means he and his team are beginning work with hundreds of PTSD patients using MDMA therapy. Therapy involves two therapists for all-day sessions and the creation of soundscapes as backgrounds. However, no religious meditation or contemplation is used during the therapy, in order avoid influencing the patient—improvement depends on the individual's "inner healing power."[77]

Here it is psychiatrists and biochemists who are interpreting the spiritual and religious imagery in the ecstatic states evoked by these drugs, and they are consciously avoiding any sort of religious interpretation. This is partly because they are not trained in the field. However, we may note that both religious and moral insights occur during this process.[78] Traditional religious authorities have thus far shown little interest in the process.

Roland Griffiths has been doing research on ecstatic states resulting from the drug psilocybin, as shown in his 2006 article "Psilocybin can occasion mystical-type experiences having substantial and sustained personal meaning and spiritual significance." His testing involved giving patients the drug in a clinical setting and evaluating them with a variety of tests, including the Hood Mysticism Scale, the States of Consciousness Questionnaire, and the Persisting Effects Questionnaire. About 33% of the participants rated the experience as the most spiritually significant event in their lives, while another 38% rated it as one of the top five most spiritually significant experiences of their lives. This was a double-blind, clinical study, the sort of research that was starting to become popular as psychiatrists and medical researchers explored ecstatic states of consciousness in clinical trials in the 1960s. He is currently doing research on the effects of psilocybin on fear of death in cancer patients.[79]

The organization MAPS (Multidisciplinary Association for Psychedelic Studies) has a broad collection of literature on psychedelic therapy on their Web site.[80] After being illegal for fifty years, we can see slow steps to examine both the healing and the ecstatic and spiritual effects of the psychedelic drugs being explored once again—though only by psychiatrists and physicians. Religious ecstasy here is being explored as a source of healing, but it has also been understood as a source of knowledge and explored from a variety of perspectives.

Ecstasy and Knowledge—The Theoretical Side

A very different explorer of states of religious ecstasy is Andrew Greeley, who writes from an unusual position as both a sociologist and a Catholic priest. He differs in his opinion of ecstasy from more mainstream sociology, which he describes as tending to dismiss ecstatic states as deviant and problematic experiences. In his book *Ecstasy: A Way of Knowing*, he notes that such states have been experienced by millions of people, and they are primarily cognitive, more about knowing than feeling.[81] Indeed, cognition is at the core of such experiences—the ecstatic sees, understands, perceives, and comprehends in a different way, for "he sees things as they really are."[82] For Greeley, ecstatic cognition is a legitimate alternative to scientific cognition.

Sometimes Greeley speaks more as a priest than as a social scientist. He notes that ecstasy does not require special divine intervention—it is the result of a "special and unusual process that goes on within the human personality."[83]

> God need not make any special intervention to become present in our lives, because in fact he is there all the time. The ecstatic interlude simply recognizes his presence.[84]

This is a cognitive religious ecstasy, which brings about a different perception of the ordinary world. However, Greeley notes that some ecstatic states are more emotional, with a passionately loving God which he describes as "the fiery, dazzling, loving, enraptured Reality that rushes in and takes possession of the human personality at the moment of ecstasy."[85]

While divine action may evoke ecstasy, so may sex and drugs. Ecstatic sex is an "act of passionate interpersonal love" rather than a hasty tension release, which is furtive, routine, anxious, or exploitative. Drug-induced ecstasy may occur when drugs trigger the state—in the same way that music or the beauties of nature may act as a trigger. He notes that there is no automatic guarantee that any drug will evoke a mystical state. However, such states should not be rejected simply because of the way that they were induced. The most important aspect of such experiences is knowledge. As Greeley states,

> Like all other forms of human knowledge, the ecstatic experience is a means whereby man understands the world of which he is a part, but it differs in being a more direct, immediate, and intuitive form of knowledge.[86]

It is one of the four major types of knowledge that Greeley argues are necessary for mankind. One is rational understanding, "common sense," also called practical or technical reason. Another type is metaphysical, dealing with ultimate issues. There is also mythopoetic knowledge, which is symbolic, and expresses ideas through images, rituals, and stories. Finally, there is mystical knowledge, which is direct and immediate, passive toward a Reality which comes rushing in. This is ecstatic knowledge, which gives direct insight into Ultimate Reality. It tends to be marked by graciousness, hopefulness, and intimacy with the Cosmos.[87] This sort of ecstatic knowledge can come in many ways, though Greeley emphasizes grace rather than meditation as its source. It is different from philosophical and ritual knowledge, his two earlier categories. He emphasizes that all of these forms of knowledge are valuable, though some tend to be neglected in the modern world.

A more recent writer who analyzes ecstatic knowledge is Ken Wilber. Wilber's work, beginning with his 1977 book *The Spectrum of Consciousness*, incorporated ecstatic states into the range of possible human states of awareness. He mapped these on a grid (with a series of sub-mappings) which he called "the spectrum of consciousness," which includes as its ecstatic states the Godhead, *ananda*, the noumenon, the void, pure reality, non-duality, cosmic consciousness, and mind. His framework is intended to be a synthesis of Eastern and Western approaches to the universe. As he notes,

> The thesis of this volume is, bluntly, that consciousness is pluridimensional, or apparently composed of many levels; that each major school of psychology, psycho-therapy, and religion is addressing a different level; that these different schools are therefore not contradictory but complementary, each approach being more-or-less correct and valid when addressing its own level. In this fashion, a true synthesis of the major approaches to consciousness can be effected… [it places] the *roots* of psychology in the fertile soil of metaphysic but without in any way harming its branches.[88]

Consciousness develops from the pre-personal to the personal to the transpersonal (understood as integrating previous levels, and as mystical awareness). Wilber has continued his research in the field of consciousness studies, and in his 2006 book *Integral Spirituality*, he describes waking, dreaming, and altered states. The altered states are divided into two groups, exogenous or induced states: which are intentionally generated from outside influences such as psychedelic drugs; hypnosis; and psycho-therapeutic techniques; and endogenous or trained states: which are intentionally generated from inside influences such as training in meditation and prayer, and religious visualization. A fourth category of states is spontaneous or peak ecstatic states, which refer to unintentional or unexpected shifts of awareness to deeper states of consciousness.[89] Wilber is interested in the role of these states in modern culture, and he refers to his work as Neo-perennial philosophy or Integral theory.

Wilber divides his research into four main phases: Wilber-1 was Romantic; Wilber-2 was based on the Great Chain of Being understood in developmental terms; Wilber-3 suggests that there are several developmental lines that progress relatively independently through the various levels of the Great Chain of Being; and Wilber-4 sets those levels and lines in the context of four quadrants of identity. His overall psychological model consists of a variety of waves, streams, selves, states, and realms. His understanding of ecstatic states is hierarchical, and people evolve through many developmental stages. His more recent work has examined the future of religion which he organizes into aspects of integral religion, questioning how religions can remain true to their central teachings while also integrating those teachings into a new framework that is inclusive of ongoing discoveries. In his 2018 book on Integral Buddhism, he argues that Buddhist teachings themselves suggest an ongoing evolution leading toward a more unified, holistic, and interconnected spirituality. Wilber describes the ways in which the tradition has been open to the continuing unfolding and expansion of its own teachings, which incorporates ecstatic religious experiences.[90]

Another researcher on ecstatic states is the psychoanalyst Dan Merkur. He has studied psychedelic states leading to mystical experiences, and in his book *The Ecstatic Imagination*, he describes about two dozen different types of ecstatic states. He takes these from the experiences of his own patients and from the available literature on psychedelic experiences. Merkur defines ecstasy as "any state of involuntary belief in the reality of the numinous." He defines the numinous as the holy or sacred, not in its intellectual or moral aspects, but rather as its sense of reality, sometimes appearing to be more real than physical reality.[91] Ecstatic states are "intrinsically religious," whether realized at the time or later on. Merkur prefers to use the term "unitive," which he considers to be broader than the terms "mysticism" or "ecstasy."

In the book, he gives a wide range of such unitive states: these include an awareness of infinite being; transcendence of the individual self; absorption of the whole of external reality within the ecstatic person; the oceanic feeling; the person radiating out realities; identification with external objects; union with nature; the self as changing or evolving through time; an immersion in the past or future; a sense of cosmic order; universal empathy and love; morality; being filled with God or cosmic power; universal life; omniscience; omnipresence; and salvation. These are not encountered as rational or theological categories, but rather as intense personal states of union. Merkur does not form them into hierarchies, but contributes them as moving toward a greater understanding of the breadth of ecstatic states.

Ecstatic and mystical states are also important in the study of esoteric knowledge. Esotericism refers to insight or understanding of inner or spiritual realities, based on a specific teaching or spiritual practice or wisdom tradition, a mystical interpretation of scriptures, or an experience of transpersonal or transcendent states. It contrasts with exoteric knowledge, which is well-known or public, and does not require hidden teachings or transformation of consciousness. Esotericism is the collective concept under which a variety of cosmological and occult teachings can be included, and Western esotericism incorporates such influences as Jewish Kabbalism, Christian Mysticism, Hermetic thought, and Gnosticism. Most understandings of Western esotericism include the "occult philosophy" of the Renaissance and its later developments. We also see Alchemy, Rosicrucianism, Christian and post-Christian Kabbalah, Theosophy, Anthroposophy and later parapsychology and transpersonal psychology. Some forms of esotericism also include Neo-Platonism, Sufi thought, and Hindu and Buddhist styles of tantras well as such practices as astrology and magic, and various related new religious movements and secret societies. As an academic field, esotericism refers to the study of alternative or marginalized religious movements or philosophies whose practitioners distinguish their own beliefs, practices, and experiences from larger, public, institutionalized religions. There is now a Western Esotericism group which is a part of the American Academy of Religion, and there are smaller academic societies, such as the American Society for Esotericism.

Esotericism is often associated with ecstatic states, for ecstasy may be both a practice to gain knowledge, and a goal to attain. Ecstatic writers from Plotinus to Jacob Boehme, William Blake, Emanuel Swedenborg, and Rudolf Steiner have emphasized the importance of spiritual or transcendent knowledge (*gnosis*), which may be described as the experiential apprehension of spiritual truths. The primary academic approach to the study of esotericism has been historical and comparative, and associated with perennialism or cross-cultural spiritual ideas, but its emphasis is on knowledge which arose from mystical and ecstatic insights.

In the area of medicine, religious ecstasy has recently been studied in a subcategory sometimes called "neurotheology." The term "neurotheology" was originally used by Aldous Huxley in his 1962 novel *Island* and has been popularized by radiologist Andrew Newberg and psychiatrist Eugene d'Aquili.[92] These scientists have studied the biological substrate of mystical experience, to determine common cross-cultural elements. They understand ecstasy to occur by the use of two methods: top-down (relaxation, meditation, and prayer) and bottom-up (excitation, dancing, and chanting). The extremes of either cause a biological "spillover effect" which brings ecstatic serenity, and they hope to be able to differentiate mysticism from psychopathology.[93] Newberg has been scanning the brains of meditating Buddhists and praying nuns to find neural correlates of mystical experience, and has noted that frontal and parietal activity decreases during these states. While the biological correlations of ecstatic states are examined, they avoid total reduction of such states to material causes.

In their book *Why God Won't Go Away*, Newberg and D'Aquili emphasize two areas of our brains that hold spatial senses: the orientation association area in the left cerebral hemisphere, which mediates the sense of self, the body's edges, and the ego location; and the orientation association area in the right hemisphere, which locates people in the world outside and gives a sense of what is not part of us.[94] Depending on whether these areas are hyperaroused, normal, or quiescent, people have vastly different perceptions of their selves and their boundaries. The various combinations of these parts of the brain, the degree of their arousal, and input from other parts of our brains and from the external environment determine perception. If the left association area is quiescent, the usual sense of a self with boundaries disappears; if the right hemisphere is quiescent, there seems to be no limit to space. When these two events occur together, people experience the sense of an endless self in an endless void, which is found in certain types of mystical experiences. On the other hand, if another part of the brain is deeply occupied with a meditative thought, a prayer, or an external object of contemplation, that content may flow into the right hemisphere and the person will feel at one with the object. While neurotheology cannot deal with theological claims, it does try to find physiological and neurological correlates for religious experiences.

It is also able to distinguish types of mystical states—Newberg and D'Aquili argue that religious experience differs in quality and effects from disease symptoms like hallucinations and hearing voices. The "impaired states" are involuntary, fragmented, incoherent, grandiose, and distressing to the person involved, who often becomes irritable as a result. On the other hand, "authentic mystical states" bring joy, peace, humility, and greater energy for life, as well as better social functioning.[95]

However, finding ecstasy in locations in the brain is reminiscent of phrenology. It was based on the idea that the brain is the organ of the mind, and that certain brain areas have localized, specific functions. More broadly, locating certain personality traits in bumps on the head was thought to explain them. The notion of "location as explanation" has been taken over by some aspects of neurobiology.

The research area of consciousness studies also interests many modern philosophers of science. They have written on the implications of the somatic nature of religious experience, pointing out the roles of the frontal lobe orientation and attention areas of the brain in the religious feelings of timelessness and spacelessness, the role of the deafferentation of the primary, egoistic circuit of the brain in the sense of egolessness reported by religious practitioners, and the role of holistic and unitary right-brain activity in the state described as pure consciousness.

As philosopher Laura Weed notes in her paper, “Religious Experience and Recent Brain Research,” neurological research reveals brain-based grounds for claiming that religious knowledge, while ineffable and non-cognitive, parallels sensory and cognitive knowledge in important respects. She terms religious knowledge “ananda,” after the Sanskrit term for bliss. Like cognitive knowledge, acquiring ananda knowledge requires instruction, extensive practice, symbol-interpreting ability, and a long-term commitment to achievement. Like sensory knowledge, ananda knowledge “is sometimes imagistic and impressionistic, often sensual or kinesthetic, often induces elevated emotional states, is difficult to put in words, and features patterns of gestalten more than atomistic bits of information.”[96] She argues that this is a new type of knowledge for Western philosophy and an important one. It can counter stress by shutting down areas of the brain associated with cognition and sense perception, and give knowledge based on different types of arousal circuits in the brain. The role of the brain, and alternative knowledge, has also become important in the field of psychiatry, with spiritual and religious experience entering its Diagnostic and Statistical Manual of Mental Disorders, the DSM, its source for insurance coverage and diagnostic legitimacy.

Modern Psychiatry and Ecstatic States

Perhaps the most well-known and respected compilation of research on consciousness, which combines both the theoretical side and the concern for healing, is the Diagnostic and Statistical Manual of Mental Disorders, accepted by the American Psychiatric Association. It includes the most recent accepted research on all mental health disorders for both children and adults. Mental health professionals use this manual when working with patients in order to be able to categorize their illnesses and find appropriate treatments, and to determine whether insurance will cover the costs. It is sometimes called the Bible for any professional who makes psychiatric diagnoses in the USA and many other countries, and its focus is on the medical model of psychopathology. Early editions of the DSM have not been sympathetic to ecstatic states. However, in 1994, there was a milestone—the DSM-IV added a new diagnostic category, "religious or spiritual problems." This meant that psychiatrists no longer had to translate religious issues into the language of psychopathology. This new category was heralded in the *New York Times*, as a sign of psychiatry opening up to religious experience.[97] While there is a new, updated version of the book, the 2013 DSM-5, most of the debate about religious experiences went on in relation to the DSM-IV.

In the chapter on "Religious or Spiritual Problems," the DSM-IV authors note that, though religious and spiritual aspects of culture have often been considered extremely important for understanding human experience, beliefs, values, and behavior, "psychiatry, in its diagnostic classification systems as well as its theory, research and practice, has tended either to ignore or pathologize the religious and spiritual dimensions of life."[98] They use the terms carefully, using the term *religiosity* to refer to "adherence to the beliefs and practices of an organized church or religious institution" and the term *spirituality* to describe "the transcendental relationship between the person and Higher Being, a quality that goes beyond a specific religious affiliation."[99] The DSM-IV notes the poor training of mental health professionals in these areas, and their difficulty in dealing with issues that involve these areas. There is often a "religiosity gap" between mental health professionals and their patients, which makes the practitioners less sympathetic to their patients, and less able to interpret their experiences. Though descriptions of problematic religious experiences have often been interpreted as signs of mental and emotional problems, as the DSM-IV notes, "some of these complaints may not be associated with psychopathology."[100] In the previous version of the DSM, the third edition, there was no category for religious and spiritual problems, and the article notes that this shows the need for a more "culturally sensitive psychiatric classification system." The combination of lack of religious training for professionals, and no clear way to understand symptoms with religious imagery, has resulted in instances of misdiagnosis; mental health professionals are not trained to distinguish between ecstasy as an illness or a revelation.

The response of the DSM-IV authors to concerns about religious experience was to do a literature search on empirical research published from 1978 to 1988 on mystical experiences, to better understand its potential to distress patients. Their computer search included variants of the word mystic, peak experience, religious, transcendent, transpersonal, spiritual, visions, and ecstasy, and they ended up with about three hundred sources. They noted the scarcity of studies in these areas, and the few that were done primarily involved psychopathological understandings of religious beliefs and images of patients. They concluded that, while "many of the leading theorists have pathologized religious and spiritual experience" it may be time for a change. This is because recent studies have found no association between religiosity and psychopathology in the non-patient population.[101] In many cases, people with religious experiences have been shown to have greater health, stability, and ability to adapt to circumstances.

The tendency to pathologize religious and spiritual experience is particularly marked in the literature of mystical experience. The DSM-IV notes that Freud reduced the "oceanic feeling" of mystics to "infantile helplessness," while people seeking mystical states showed a "regression to limitless narcissism." The 1976 report from the Group for the Advancement of Psychiatry entitled "Mysticism: Spiritual Quest or Mental Disorder?" interpreted mysticism in a similar way, as an escape and a projection of a primitive infantile state.

However, studies have found that people reporting mystical experiences scored lower on psychopathology scales and higher on measures of psychological well-being than their control subjects.[102] They cite a variety of clinical psychologists and theorists who have challenged the trend to pathologize religious experience, viewing mystical experiences as a sign of health and an agent of transformation. The theorists that they cite are classical writers on mysticism, such as William James; many of the sources that they used were about a century old, which says much about the limited availability of modern sources.

It would be useful for psychoanalysts and psychologists to have more training in understanding ecstatic and mystical states; according to one survey, psychologists estimated that at least one in six of their patients had problems that involved religion or spirituality, and 72% of psychologists queried stated that they had addressed religious or spiritual issues in treatment.[103] In 1990, the American Psychiatric Association Committee on Religion and Psychiatry organized guidelines for psychiatrists, noting that they should respect the faith and values of their patients, and not impose their own "religious, antireligious, or ideologic systems of belief on their patients."[104] Indeed, they note that with the range of religions of their international patients, and the rise of new religious movements, mental health professionals must now deal with such areas as unexpected mystical experience, near-death experience, spiritual emergency (such as kundalini awakening), and shamanic initiatory crisis. As an example of modern problems, the study states that according to both medical research and a 1982 Gallup poll, approximately one-third of all individuals who have had a close encounter with death have had a near-death experience. These tend to result in an increased appreciation for life, concern for others, and an "overall shift towards universalistic spiritual values regardless of previous religious affiliations or lack thereof."[105] Yet these experiences have long been ignored or interpreted as pathological by health professionals.

There tend to be strong differences in religious views between psychiatrists and the general population; the DSM-IV cites the Gallup Poll from 1986, in which approximately 50% of the psychiatrists described themselves as atheists or agnostics, contrasting with between 1% to 5% of the general population who described themselves in that way.[106] The DSM-IV authors noted that psychiatrists, psychologists, social workers, and marriage and family counselors did tend to endorse the item that said "Seek a spiritual understanding of the universe and one's place in it." They suggested that spiritual interests are often unexpressed because of the secular presuppositions of professional education and practice. Such a "spiritual humanism" could allow an area of commonality between clinicians and the public.

The DSM-IV authors gave five possibilities for future directions of response to religious and ecstatic states. Clinicians could ignore religious and spiritual issues that are not due to a mental disorder, to avoid the "psychiatric imperialism" which claims that psychiatrists have greater expertise than they actually possess. However, this leaves the field open to misdiagnosis, mistreatment, a poor reputation, and inadequate research and theory. Another option would be to include religious and spiritual issues under the existing diagnostic categories for mental disorders. For instance, a mystical experience could be diagnosed as a regressive state. However, mystical states tend to lack both stressors (specific life stresses which cause the reaction) and specific adjustment problems (difficulty in returning to a less regressed or ordinary state). A third option would be to include these problems under other conditions that are not specific mental disorders. But if they were categorized under identity problems or phase-of-life problems (rather than as "mental disorders"), they would lack the necessary developmental framework for understanding them, and this would not help with diagnosis or treatment.

A fourth option would be to create a new diagnostic category as an area needing further study. This was considered as a possibility, but the authors ended up supporting a fifth option, the creation of a separate new diagnostic category entitled "psychoreligious or psychospiritual problem." This would be a non-mental-disorder category, which could be an appropriate response to either problems associated with institutional religions or with mystical states (including near-death experience). The categorization is still up for debate. Another area of debate is the distinction between "pure" religious/spiritual problems, which should be addressed by religious professionals or spiritual teachers, and psychoreligious and psychospiritual problems, which would require mental health professionals for diagnosis and cure. All of these attempts are trying to move psychiatry in a new direction of sympathy with religious and mystical experiences.

The current DSM-5 version has added more categories, but it has done little to further clarify ecstatic and mystical states. The DSM-5 includes issues of religion and spirituality, listing them as part of either pathology or culture. Religion as a cultural derivative reflects the human aspect of religion, and the DSM emphasizes that it does not speculate on a client's perception of divine actions as transcendent. It attempts to discern if a client's experience reflects pathology or is based on a cultural assumption. It accepts the V code from the DSM-IV:

> V62.89 Religious or Spiritual Problem: This category can be used when the focus of clinical attention is a religious or spiritual problem. Examples include distressing experiences that involve loss or questioning of faith, problems associated with conversion to a new faith, or questioning of spiritual values that may not necessarily be related to an organized church or religious institution.

This category includes such wide-ranging issues as Mystical experience, New Religious Movements and Cult membership, Psychic opening, Visionary experience, Kundalini awakening, Near-death experience, Possession experience, Shamanic crisis, Loss of faith, Alien encounters, Terminal & life-threatening illness, and Changes in religious membership, practices, and beliefs.[107] A variety of articles have been written debating the usefulness of the DSM-5 for analyzing religious experiences from non-Western cultures.[108]

The psychoanalyst Sudhir Kakar finds the debate over the legitimacy of ecstatic and mystical states to be part of a broader tension between the romantic and rational approaches to life. He notes that the rational is currently dominant, both in ordinary psychology and in academia, with a suspicion of higher powers, and a desire to show that all gods have clay feet. However, he argues that we are poised to witness a resurgence of romantic vision, in which the culture returns to valuing union with a higher power or Spirit. From the spiritual perspective, people are not limited to the enduring separation between humankind and the universe found in the rationalist perspective. Instead, an intimate connection can be found between them.[109] Kakar notes that the most likely locale for return to a romantic perspective is the West:

> I must note here that, as a broader intellectual and social current, the romantic resurgence is more characteristic of Western societies that of most non-Western ones, where the romantic vision never lost its ascendancy over the rational... in the non-Western world, the movement is in an opposite direction as educated professionals enthusiastically embrace the rationalist vision of the Enlightenment.[110]

Kakar uses as evidence the recognition of the importance of altruism in the neurosciences and evolutionary psychology, the quest for authenticity (which he links to the romanticism of Rousseau), and the modern interest in self-transcendence, ranging from a loving connection to an object to the mystical union of the saints. He wishes to add to psychology a "hermeneutics of idealization" which examines the potentialities of the psyche, its possible extension into realms of spirit, as opposed to its current contentment with Freud's goal of replacing hysterical misery with common unhappiness.[111] He suggests that studying the psychological influences on ecstatic states, such as primal fantasies, would be a good direction for psychological research.

If psychiatry and psychoanalysis can move from their former condemnation of ecstatic states as infantile and regressive, to being open to the possibility that they may give special and valuable forms of knowledge, perhaps other institutions in the modern West can follow.

We have explored the role of ecstasy from many perspectives. In the last chapter, we shall explore some of the problems that the current approaches to ecstasy have caused, and some possible ways of dealing with them.

Notes

1. While Das Heilige is usually translated into English as *The Idea of the Holy*, simply *The Holy* would be a much better translation. His point in the book is that we should study the experience rather than the idea.
2. Rudolf Otto, *The Idea of the Holy (Das Heilige): An Inquiry into the Non-rational Factor in the Idea of the Divine and Its Relation to the Rational*, trans. John W. Harvey (New York: Oxford University Press, 1958 (1923)), p. 12.
3. Ibid., p. 5.
4. Ibid., pp. 32–35.
5. Ibid., p. 29.
6. A centennial issue of the *Varieties* was recently published, with commentary by modern scholars, and the American Academy of Religions recently had several panels which focused upon James' ideas.
7. William James, *The Varieties of Religious Experience* (New York: Collier Books, 1974 (1902)), p. 24.
8. Ibid., p. 300.
9. Ibid., p. 305.
10. Ibid.
11. Ibid.
12. Ibid., p. 318.
13. Ibid., p. 323.
14. Ibid., p. 324.
15. Ibid., p. 335.
16. Ibid., p. 273.
17. C. G. Jung, *Collected Works of CG Jung, Volume 11, Psychology and Religion, West and East* (New York: Pantheon Books, 1958), p. 544.
18. From *Jung Speaking: Interviews and Encounters*, ed. McGuire and Hull, 1977, cited in *Carl Jung and Soul Psychology*, ed. Karen Gibson et al. (New York: Harrington Park Press, 1991), p. 3.
19. C. G. Jung, *Two Essays on Analytical Psychology*, ibid., p. 238.
20. C. G. Jung, *The Integration of the Personality* (New York: Farrar and Rinehart, 1939), p. 91.
21. C. G. Jung, *Psychology and Religion*, ibid., p. 90.
22. C. G. Jung, *Collected Works of CG Jung, Volume 6, Psychological Types* (London: K. Paul, Trench and Trubner Co., 1923), p. 325.
23. Paul Tillich, "Religion as a Dimension in Man's Spiritual Life," reprinted in *Phenomenology of Religion: Eight Modern Descriptions of the Essence of Religion*, ed. Joseph Dabney Bettis (New York: Harper and Row, 1969), p. 174. The article was initially printed in Paul Tillich, *Man's Right to Knowledge* (Columbia University Press, 1954).

24. Ibid., p. 177.
25. Ibid.
26. Ibid., p. 178.
27. Mircea Eliade, *The Sacred and the Profane: The Nature of Religion* (New York: Harcourt, Brace and World, 1959), p. 13.
28. Eliade, Sacred, p. 26.
29. Eliade, Sacred, p. 30.
30. Mircea Eliade, *Yoga: Immortality and Freedom*, trans. Willard. R. (Princeton: Princeton University Press, Bollingen Series NVI, 1973 (1969) Trask), p. 320.
31. Ibid., p. 70.
32. Cited in Ernest Kurtz, *Not God—A History of Alcoholics Anonymous* (Hazelden, 1979), pp. 19–20.
33. Cited in Jalaja, ibid., p. 97.
34. Christina Grof, *The Thirst for Wholeness: Attachment, Addiction, and the Spiritual Path* (Harper One, 1994), p. 10.
35. C. G. Jung, "Letter to Bill Wilson," reprinted in *ReVision* 10, no. 2 (1987), p. 21.
36. Grof, ibid., p. 220.
37. Jung, ibid., p. 21.
38. J. D. McPeake, B. P. Kennedy, S. M. Gordon, "Altered States of Consciousness Therapy: A Missing Component in Alcoholic and Drug Rehabilitation Treatment," *Journal of Substance Abuse Treatment* 8, no. 1–2 (1991): 76.
39. Robert A. Johnson, *Ecstasy: Understanding the Psychology of Joy* (San Francisco: Harper and Row), p. vii.
40. Ibid., p. vii.
41. Jay Michaels, "Meditation and Drugs," online at http://realitysandwich.com/60367/meditation_and_drugs/.
42. Ibid.
43. Jalaja Bonheim, *The Hunger for Ecstasy: Fulfilling the Soul's Need for Passion and Intimacy* (New York: Rodale Press, 2001), p. 1.
44. Ibid., p. 2.
45. Barbara Ehrenreich, *Dancing in the Streets: A History of Collective Joy* (New York: Henry Holt and Company, 2007), p. 174.
46. Ibid., pp. 223–225.
47. Ibid., p. 250.
48. Cited in Ehrenreich, ibid., p. 250.
49. Ibid., p. 260.
50. It was a popular phrase, lasting for over twenty years (from 1980–2001). The current theme is "Army Strong."

51. Abraham Maslow, *Religions, Values and Peak Experiences* (New York: The Viking Press, 1973 (1964)), p. 19.
52. Ibid., p. 28.
53. Abraham Maslow, "Religious Aspects of Peak-Experiences," in *Personality and Religion*, ed. W. A. Sadler (New York: Harper & Row, 1970), p. 19.
54. Maslow, *Religions, Values and Peak Experiences*, pp. 60–67.
55. Ibid., p. viii.
56. Ibid., pp. 92–94.
57. Ibid., p. 104.
58. Ibid., p. 32.
59. Ibid., p. 19.
60. Abraham Maslow, *The Farther Reaches of Human Nature* (New York, 1971), p. 269.
61. Rebecca Webber, "Big Moments: Chasing Peak Experiences," *Psychology Today* 43, no. 5 (September/October 2010): 64.
62. Cited in Webber, ibid., p. 65.
63. Ibid., p. 69.
64. Stanislav Grof, "The Consciousness Revolution: New Perspectives in Psychiatry, Psychology, and Psychotherapy," p. 3, lecture online at http://www.stanislavgrof.com/wp-content/uploads/pdf/Moscow_ITA_Lecture.2010.pdf.
65. Stanislav Grof, *Realms of the Human Unconscious: Observations from LSD Research* (New York: Dutton, 1976), p. 106.
66. Stanislav Grof, *The Holotropic Mind: The Three Levels of Human Consciousness and How They Shape Our Lives* (Harper San Francisco, 1990), p. 180.
67. Stanislav Grof, *The Holotropic Mind: The Three Levels of Human Consciousness and How They Shape Our Lives* (San Francisco: HarperSanFrancisco, 1990), p. 16.
68. Stanislav Grof, *Realms of the Human Unconscious: Observations from LSD Research* (New York: E. P. Dutton, 1976), p. 34.
69. Grof, *Realms*, pp. 156–157, for a general outline. These are described in greater detail throughout the book.
70. Grof, *The Holotropic Mind*, ibid., p. 63.
71. Ibid., p. 202.
72. R. E. L Masters and Jean Houston, *The Varieties of Psychedelic Experience* (New York: Dell Publishing, 1966), p. 5.
73. Ibid., pp. 142–144. However, these levels are described throughout the book.
74. Ibid., p. 308.

75. EMDR or Eye Movement Desensitization and Reprocessing is another current treatment for PTSD along with conventional psychopharmacology, but the latter only addresses some symptoms. In Prolonged Exposure, the patient gradually learns to tolerate traumatic memories and experiences that may be triggering for them, such as being in certain places and circumstance. EMDR is similar in that it also involves retelling the traumatic narrative, but it is combined with techniques that aim to reduce sympathetic arousal and implement bilateral sensory input. The goal of therapy is to help the individual reintegrate traumatic material.
76. See Michael C. Mithoefer et al., "The Safety and Efficacy of ±3, 4-mthylenedioxy-methamphetamine-Assisted Psychotherapy in Subjects with Chronic, Treatment-Resistant Post-traumatic Stress Disorder: The First Randomized Controlled Pilot Study," *Journal of Psychopharmacology*, 25, no. 4 (April 2011): 439–452. Also "Landmark Study Shows Long-Term Benefits of MDMA-Assisted Psychotherapy" at http://www.prnewswire.com/news-releases/landmark-study-shows-long-term-benefits-of-mdma-assisted-psychotherapy-for-people-suffering-from-ptsd-180089391.html.
77. Personal communication, with Dr. Michael Mithoefer and also Dr. Yevgeniy Gelfand, who will be doing therapy for the 2018 trials.
78. I addressed this issue in June McDaniel, "Strengthening the Moral Compass: The Effects of MDMA ('Ecstasy') Therapy on Moral and Spiritual Development," *Journal of Pastoral Psychology* 6, no. 6 (2017): 721–741, https://doi.org/10.1007/s11089-017-0789-6. In interview, psychiatrists guiding patients during MDMA therapy have stated that they wish to avoid "leading" the patients with religious imagery, which they were not trained to use, and that the "natural healing energy" of the body would be the focus of the therapy.
79. For Roland Griffiths' work, see R. R. Griffiths, W. A. Richards, U. McCann, and R. Jesse, "Psilocybin can Occasion Mystical-Type Experiences Having Substantial and Sustained Personal Meaning and Spiritual Significance," *Journal of Psychopharmacology* (May 2006), https://doi.org/10.1007/s00213-006-0457-5. Also, from the National Institute of Health, Roland R. Griffiths, William A. Richards, Matthew W. Johnson, Una D. McCann, and Robert Jesse, "Mystical-Type Experiences Occasioned by Psilocybin Mediate the Attribution of Personal Meaning and Spiritual Significance 14 Months Later," *Journal of Psychopharmacology* 22, no. 6 (August 2008): 621–632, published online, July 1, 2008, https://doi.org/10.1177/0269881108094300.
80. See www.maps.org, which has a collection of relevant research papers.
81. Andrew M. Greeley, *Ecstasy: A Way of Knowing* (Englewood Cliffs: Prentice-Hall, 1974), p. 4.

82. Ibid., p. 7.
83. Ibid., p. 56.
84. Ibid., p. 68.
85. Ibid., p. 70.
86. Ibid., p. 57.
87. Ibid., pp. 60–61.
88. Ken Wilber, *The Spectrum of Consciousness* (Wheaton: The Theosophical Publishing House, 1985 (1977)), p. 11.
89. Ken Wilber, *Integral Spirituality: A Startling New Role for Religion in the Modern and Postmodern World* (Shambhala, 2006).
90. Ken Wilber, *Integral Buddhism and the Future of Spirituality* (Shambhala Publishers, 2018). He describes the processes of Integral Buddhism (or Integral Spirituality in general) including Waking Up (experiencing deeper states of consciousness), Growing Up (organizing the structures of deeper consciousness), and Cleaning Up (shadow work, dealing with lingering psychological problems).
91. Dan Merkur, *The Ecstatic Imagination: Psychedelic Experiences and the Psychoanalysis of Self-Actualization* (Albany: State University of New York Press, 1998), p. 92.
92. See John Horgan, *Rational Mysticism: Dispatches from the Border Between Science and Spirituality* (Boston: Houghton Mifflin Co., 2003), p. 73.
93. Horgan, ibid., p. 74.
94. Andrew Newberg, Eugene D'Aquili, and Vince Rause, *Why God Won't Go Away: Brain Science and the Biology of Belief* (New York: Ballantine Books, 2001).
95. Ibid., pp. 111–113.
96. Laura Weed, "Religious Experience and Recent Brain Research," unpublished paper, given at the 21st World Congress of Philosophy, 2003.
97. Russell Shorto, *Saints and Madmen: Psychiatry Opens Its Doors to Religion* (New York: Henry Holt, 1999), p. 17.
98. Francis G. Lu, David Lukoff, and Robert Turner, "Religious or Spiritual Problems," in *DSM-IV Sourcebook*, ed. Thomas A. Widiger et al. (Washington, DC: American Psychiatric Association, 1997), Chapter 56, p. 1001.
99. Ibid., p. 1002.
100. Ibid.
101. Ibid., p. 1004.
102. Ibid. They list four studies by Caird, Hood, Spanos, and Moretti on this topic. The assumptions may be shown in the title of the first cited, by D. Caird, "Religion and Personality: Are Mystics Introverted, Neurotic or Psychotic?".

103. Ibid., p. 1005.
104. Ibid.
105. Ibid., p. 1007, primarily citing the work of Kenneth Ring.
106. Ibid., p. 1008.
107. See *Religious and Spiritual Issues in Psychiatric Diagnosis: A Research Agenda for DSM-V*, at https://religiousandspiritualproblems.blogspot.com/.
108. See, for example, Romara Delmonte B.Sc., Giancarlo Lucchetti MD, Ph.D., Alexander Moreira- Almeida MD, Ph.D. & Miguel Farias B.Sc., MA, DPhil, "Can the DSM-5 Differentiate Between Nonpathological Possession and Dissociative Identity Disorder? A Case Study from An Afro- Brazilian Religion," *Journal of Trauma & Dissociation* 17, no. 3 (2017): 322–337, online at http://dx.doi.org/10.1080/15299732.2015.1103351. The answer to their question is, Not very well.
109. Sudhir Kakar, *Mad and Divine: Spirit and Psyche in the Modern World* (Chicago: University of Chicago Press, 2009), p. 1.
110. Ibid., p. 2.
111. Ibid., p. 6.

CHAPTER 10

Conclusions: Can We Go Beyond Criminalizing, Pathologizing, and Trivializing? Or, the Problems of Shooting Yourself in the Foot

Ecstatic experience has undergone many attacks in both world religions and the academic fields which study them. When it is understood as foreign or threatening to the current social values, mystics become heretics, criminals who are threatening the social order. Thus, we have witch-burnings and the jailing and killing of visionaries. When it is examined through the dominant psychological approaches, it becomes a form of illness, whether hallucination, narcissistic inflation, responses to trauma, or dissociative identity disorder. With the models of brain pathologies, ecstatic experience is a sign of abnormal brain activity, of malfunctions or anomalies. When it is examined through the lenses of modern Religious Studies, the experiences themselves are much less valuable that how people use them, and they are trivialized as only a background for the battles over power and status which occur in the social world around religious founders, ecstatics, and prophets.

The "wreck of the good ship Eliade," as it was phrased at the perennialism session at the 2017 American Academy of Religions meeting, has stranded the field of History of Religions in area studies. It was a ship sunk by its own crew, to avoid the broader waters and "big questions" that the field of philosophy has also chosen to avoid. But these are issues that interest students, and the drumbeats of emphasis on power hierarchies, postmodernism, Foucault, Asad, and post-colonialism (the cluster called "Pomofoco" by Jacques Berlinerblau), has guaranteed the field will be a shrinking one, a subset of political science or economics rather than a field of its own.[1]

J. McDaniel, *Lost Ecstasy*, Interdisciplinary Approaches to the Study of Mysticism, https://doi.org/10.1007/978-3-319-92771-8_10

This is the aspect of the field that is "shooting itself in the foot." If we return to our shipwreck of a field, it has jettisoned its cargo of religious knowledge because it is no longer valued. Much of the field consists of saying how the cargo of myths, symbols, visions, and value systems was worthless in the first place, and our job is to explain its problems of power and privilege, its inaccuracies, and its inadequacies. When the job of a field is no longer passing on knowledge, but rather criticizing it, that knowledge becomes less valuable in popular perception. Why should students want to study a field which its own professors do not respect, and whose major thinkers who formed the field are an embarrassment? And for the future, how can it be defended from retrenching by administrators who want smaller numbers of departments in the Humanities?

When toolboxes get smaller, less can be accomplished. When scholars denigrate their own objects of study, or constrict and redefine their fields to arguments over power and dominance, they lose their audience. If the field of mathematics were to be continually critical of belief in numbers, avoiding discussion of equations or undermining their importance and usefulness, how many people would want to study mathematics? Can physicians who do not believe in the value of drug therapies or who think surgery or blood transfusions are never desirable be effective healers? Can an administrator who believes that the environment is there to be exploited for wealth and personal gain be an effective leader of an Environmental Protection Agency? It is a postmodern tendency to cast doubt on one's own field and see scholarship as an academic game to gain authority or power. But the process of deconstruction often lacks its necessary reintegration.

Paul Ricoeur's writing, which has inspired many modern scholars with his hermeneutic of suspicion, tends to be used in part rather than in toto. There was more to Ricoeur than skepticism, indeed, as he phrased it, "beyond the desert of criticism, we wish to be called again."[2] The hermeneutic of suspicion was only a stage, and it is certainly useful to overcome the narrowness of the first stage of blind faith or first naivete. The first naivete is a precritical stage, where what is taught by parents and authority figures is understood as true without question. Some critical distance is needed to question these traditional teachings, examining the distinctions between ancient and modern worldviews, looking for biases, agendas, exaggerations, and distortions.

But the hermeneutic of suspicion is not the endpoint of analysis. It is followed by the second naivete, the level of integration or individual faith. It is post-critical or meta-critical, and it deliberately avoids having critical methodology become an ideology or faith of its own. As Ricoeur writes, “The dissolution of the myth as explanation is the necessary way to the restoration of the myth as symbol. Thus, the time of restoration is not a different time from that of criticism; we are in every way children of criticism, and we seek to go beyond criticism by means of criticism, by a criticism that is no longer reductive but restorative.”[3] Indeed, he notes that “The second naivete aims to be the postcritical equivalent of the precritical hierophany.”[4]

We do not see Ricoeur’s interest in reconstruction, refiguration, and the return to the value of symbolism, reflected in the modern emphases within Religious Studies. This is where skepticism and doubt were intended to lead, to a new integration of ideas which did not only emphasize attack, but also emphasize creativity. While we continually hear of the hermeneutic of suspicion, what has happened to the hermeneutic of integration, our post-critical hierophany?

In the constructivist school of mystical interpretation, each form of mysticism is separate, and there is no common deeper state that can be shared or compared. One reason for the lack of emphasis on common interests and themes between experiences in world religions is said to be that Western scholars (primarily Protestant ones) project their own ideas and values on other cultures. While the focus on absolute separation of Western philosophy and Theology from foreign religions is said to show respect for their systems of belief and practice, thus avoiding ways that their members might be similar to Westerners, it is an open question whether this distancing from the religions of others actually shows respect for them. Do we respect a culture by disrespecting its values? Many traditional religious cultures value ideas that postmodernism would not value. These include introspection, universal claims, beliefs in the supernatural, and the divine. Do we show our respect for a culture by dismissing its ideals and making it alien from us?

This is why Religious Studies could use a “big tent” approach, allowing for a variety of perspectives. At times, it resembles political parties who expel their moderates in order to focus on the base, still fighting the conflicts of the European Enlightenment between skepticism and Theology. The study of mystical and ecstatic experience needs to be more than an echo of political science, economic theory, game theory, neurobiological problems “deemed religious” and cultures enslaved to “lineages of power.” Considering its importance in the history of ideas and the History of Religions, religious ecstasy deserves better.

Leaving religious experience to theologians means that only a narrow band of that experience can be analyzed—most will be rejected as not doctrinal. Each Theology has a narrow range of acceptable supernatural visions and abilities and rejects the others as unworthy of study. When we leave religious ecstasy to the theologians, very few examples can be studied.

As we have seen in this book, there are many people for whom intense ecstatic experience is an important aspect of life. If religions will not offer a situation in which such states can be attained and directed, people will find them elsewhere. While some ecstatic states are useful for both the person and the society around him or her (such as intense artistic or scientific creativity), we see the drive toward intense experiences more often in areas like pleasure (shown in the sex, drugs, and music chapter), or in danger (extreme sports, extreme fighting, the joys of war and violence, also shown previously). The desire for intense experiences cannot be erased merely by condemning it, or ignoring it, or by criticizing it.

It is useful for us to explore and understand how ecstatic experiences work, to create more empathic frameworks or methodologies through which to understand them. Theology can discuss the sectarian origins of a state of religious ecstasy, and sociology can discuss how people in these states have been understood by the societies around them. But much of the writing that examines ecstatic and mystical states is advocacy journalism, in favor of it or against it. These days, we find the majority of academic literature to be on the opposition side, and this book has examined some of their condemnations of ecstatic experience.

Our understandings of ecstatic states are based on the constructs and models that we use, and we have noted three major models of condemnation. If ecstatic states of consciousness are believed to be danger to public order, something corrupting the youth rather than a means of gaining healing and insight, they will be criminalized. The criminalization of ecstatic states has ranged from the medieval wars against heresy, in which alternative religions and their prophets were understood as dangerous to society, to the modern drug wars, in which the psychedelics were understood to evoke states of political rebellion, sexual freedom, addiction, madness, and heathen nirvana. If ecstatic states are interpreted according to a disease model, resulting from such problems as hereditary abnormalities, malnutrition, poverty, brain dysfunction, and childhood trauma, they will be pathologized as diseases seeking a cure and understood as problems seeking a solution. If ecstatic states are viewed as misunderstandings, logical mistakes due to ignorance, gullibility, misinterpretation of texts, and linguistic ambiguities, as mere distractions from fights over power, they will be trivialized as errors which can be fixed by improved education and increasing maturity. Whether criminalized, pathologized, or trivialized and patronized, ecstasy has become an unacceptable subject for academic writing.

Cultural condemnation may force experiences of religious ecstasy to be expressed in other areas than traditional religion. As Victoria Nelson writes in her book *The Secret Life of Puppets*, ecstatic states cannot be seriously believed in modern religion, as religious ecstasy is a "quaint construct," a part of the Platonic cosmos dead to all but scholars. Modern people do not believe in the possibility of religious ecstasy, for it has been expelled from the main currents of Western intellectual culture. Thus, ecstatic states can now be expressed only in areas like the visual arts and in fantasy literature, where one can suspend disbelief:

> In the current Aristotelian age the transcendental has been forced underground, where it has found a distorted outlet outside the recognized boundaries of religious expression. As members of a secular society in which the cult of art has supplanted scripture and direct revelation, we turn to works of the imagination to learn how our living desire to believe in a transcendent reality has survived outside our conscious awareness.[5]

As Nelson discusses, modern culture represses spirituality, bracketing deeper questions of life, and spirituality moves into popular culture and is shown in fantasy, science fiction, and films. Art becomes the source of religious meaning, through myth, horror, and thoughtful robots and cyborgs. The culture as a whole suffers from alienation born of scientism, the belief that the scientific method has all answers to all questions, a one-sided view of life that emphasizes rationalism at the expense of mysticism. It is her belief that the cultural pendulum has swung toward rationalism for three hundred years and has reached its zenith. Now, it begins its swings back toward spirit and gnosis. But the goal is balance—cultures need both Plato and Aristotle, idealism and empiricism, and these extremes should not be trying to wipe each other out. But for now, the art forms of the fantastic may be the closest that modern secularists can get to religious experience.[6]

As a historian of religions and an Indologist, I am more familiar with cultures which value ecstatic states, and they have religious specialists, philosophers, and contemplatives who examine them in great detail. In Hindu bhakti, folk religion, and Shakta and Shaiva tantra, there is discernment between accepted and unaccepted forms of ecstasy, possession by gods or ancestors, visions that are divine and visions that are hallucinations, divine madness or ordinary madness. There are ways of testing ecstatic states, which are routinely used to distinguish between their types. These ways are neither new nor post-colonial. One need only read biographies of saints like Caitanya Mahaprabhu to see that the attempts to evaluate his ecstatic states were culturally important in the fifteenth century CE. Such interest and forms of testing have continued among my modern Bengali Hindu informants. While we have seen the claim that an interest in religious experience in Eastern religions is recent, Hindu bhakti has been emphasizing ecstatic and visionary experiences for at least a thousand years, and yoga philosophy has analyzed the variety of possible ecstatic states for much longer. Stories and descriptions of such experiences have long been available in Indian oral traditions, saint biographies, devotional poems, and yogic texts. But these have been ignored by many Western scholars more interested in pragmatic concerns. As one Hindu priest said in recent fieldwork, in order to communicate with foreigners, he describes himself as "in the God business." He gains more respect this way.

In the West, people have often had to construct their own understandings of ecstatic states. It is one of the bases of the New Age and Spirituality movements, whether as nature religion, shamanism, or visions of light interpreted as angels or guardian spirits. In her book *The New American Spirituality: A Seeker's Guide*, Elizabeth Lesser (a co-founder of Omega Institute) describes the longing for such ecstatic states as a spiritual hunger, which can never quite be satisfied by substitutes:

> Neither a feeling nor a thought, it is more like a gravitational pull in the direction of wholeness, enlightenment, truth- what some call God. There are some people who know from an early age how to follow their heart's longing with grace and sureness. Others feel it as strongly and fill it with anything they can to dull the longing. Alcohol, drugs, materialism, work- many of our excesses can be traced to the spiritual longing that dwells in our hearts.[7]

Such spiritual calls or "holy longing" have been described from Biblical prophets to local shamans; we see this longing in Augustine's "God-shaped hole in the heart." It is a part of the pendulum that swings between religious experience and social organization, the Holy Spirit and the institutional Church, or in India between *bhakti* and *dharma*.

That pendulum can even be found in academia, though Western scholars rarely speak publically of the religious experiences that drew them to study religion. In the field of South Asian religion, many researchers went to study with gurus before they went to graduate schools in the field and performed meditation and other ritual practices. However, they soon learned that such topics could not be discussed easily in an academic environment. A person with such religious faith and experience could be accused of "going native," of showing too much sympathy to informants, of being biased. The false dichotomy of "curator vs critic" has created a Manichaean dualism, where it is difficult to include a central area of empathic exploration. The tension between being a critic and a believer is found in the study of world religions generally today and debated as part of the participant-observer question. Indeed, a recent novel has as its narrator a scholar working on a treatise on medieval mysticism, and he satirizes this problem,

> I personally have no belief or faith whatsoever in anything supernatural, or even transcendent for that matter, and so I thought myself eminently qualified for the difficult task of elucidating mystic states.[8]

In yet another example of this approach, while Jeffrey Kripal writes sympathetically of the sacred in his book *Authors of the Impossible*, he too avoids the issue of belief: "Indeed, as a professional scholar of religion, I consider it my job not to believe, and I take that professional commitment very seriously."[9] Yet, Kripal also notes how strange it is that academia avoids all discussion of religious ecstasy and the paranormal. He states that religious experience has been eclipsed in the study of religion, and notes "the odd conflation" of the *sui generis* nature of the sacred and the believer's perspective, "as if taking the sacred seriously is equivalent to surrendering one's intellect and critical faculties to the faith claims of the religious tradition." He too notes that by reducing claims of religious experience to material causes, "the field has, in effect, denied its own subject matter."[10] He describes the study of altered states of consciousness and reports of miraculous events:

> Whereas such marvels are vociferously denied (or simply ignored) in the halls of academic respectability, they are enthusiastically embraced in

> contemporary fiction, film, and fantasy. We are obviously fascinated by such things and will pay billions of dollars for their special display, and yet we will not talk about them, not at least in any serious and sustained professional way. Popular culture is our mysticism. The public realm is our esoteric realm.[11]

Religious ecstasy may be found as fantasy in art and literature and in the modern movements of the New Age and the spirituality movements. As the New Age movement has declined, it has left in its wake a value on "spirituality," which usually includes a search for authentic experience and practice of spiritual disciplines. These are widely varied; in some areas, there are Christian spirituality groups, with centering prayer, labyrinths, pilgrimage to holy sites, and women's spirituality and renewal group retreats. In other areas, we find Buddhist meditation, yoga, vegetarianism, and eco-spirituality. Shamanism has been redefined as a technique of self-actualization, a way to develop individual mythologies, gain a spirit guide, discover new archetypes, and find new states of consciousness.[12] Some people have been searching for truth through altered states of consciousness, using virtual reality headsets, computer simulations, control of brainwaves, music controlled by "techno-shamans," drugs, dance, drumming, and the meditative practices of other cultures.[13] We see this concern discussed by Heelas and Woodhead, in the opposition of formal institutional religion and spirituality, and by Partridge, in the issue of re-enchantment, especially through pagan and occultist spirituality.[14]

Because it does not fit in with "modern" ideas, ecstasy has been associated with a return to a more "primitive" or romantic approaches to life. In her *Primitive Passions: Men, Women and the Quest for Ecstasy*, Marianna Torgovnick examines post-World War I literature and art and their seeking after oceanic and ecstatic states. She includes D. H. Lawrence, Andre Gide, Carl Jung, and Georgia O'Keefe, among others, and notes the "primitive" is equated with reverence for nature, sexual joy, and a deep, mystical connection with the universe. While the term can refer to prehistoric societies and exotic histories that are "simple and crude," it also brings associations of ecstatic dancing, drumming, and initiation. Primitivism refers to both interest in or borrowing from indigenous people and also a desire for insight into one's origins, "to reinhabit core experiences." It has been repressed, distorted, and idealized in the twentieth century; Torgovnick notes the anthropologist Bronislaw Malinowski's theory of the primitive "nirvana impulse" in the Trobriand

Islands as an example of this distortion. Today, it is found in the fascinations with shamanism and Native American religion, pagan bonfires, piercing rituals, and various forms of return to nature. The author notes that the quest for ecstasy is as much a part of Western fears and desires as it is a part of the peoples of the forest and the desert.[15]

One reason for the neglect of ecstatic and religious consciousness as a legitimate area of study has been suggested by Alan Wallace. He calls it "the taboo of subjectivity," which he describes as due to the rise of science (especially neuroscience) as the primary way to understand human experience. He asks, "Can science provide an adequate view of the entire natural world that includes only objective phenomena, while excluding the subjective phenomenon of consciousness altogether?"[16]

He dates the "marginalization of the mind" to the latter half of the twentieth century, when cognitive psychology adopted a view of the universe in which it was compared to a giant computer, with consciousness subject to the laws of physics. Mental events lost their status as real entities, becoming mere epiphenomena of brain activity and propensities for behavior.[17] He cites the behaviorist John Watson, who wrote in 1913 that "The time has come when psychology must discard all reference to consciousness" and who noted that any belief in the existence of consciousness was due to ancient superstition and magic.[18] Wallace cites a number of philosophers and psychologists who echo the view that mental states do not exist, especially those within the philosophical school of eliminative materialism. He states that this is awkward, as the mind is "the fundamental instrument of scientific inquiry," and these are people who greatly value science. Yet for them, subjective experiences are "mere waste, equivalent to nothing at all."[19]

Thus, as Wallace observes, the exploration of subjective or inner experience has two taboos, one against scientific inquiry into subjective mental phenomena and another against having personal perspectives taint any research. He describes a 1994 conference sponsored by the Royal Society in London, which concluded that "science understands none of the central aspects of consciousness- what it is, how it evolved, how it is generated by the brain, or even what it is for."[20] Yet, its participants refused any suggestion that a first-person perspective might be useful for such exploration. Wallace suggests that part of the issue is that researchers have lost interest in the topic, part is the focus on positivism, part is an inability to determine the mechanisms involved, and part is the general "problem of privacy." He also describes several "metaphysical" objections, such as the desire for an independence of subject and object which is difficult or impossible in the study of consciousness, and the view that mental processes are unconscious and inaccessible because they are brain states.[21] Thus, consciousness is either excluded from study or described through a "neuromythology" which explains very little. The taboo of subjectivity limits the possible methodologies for the study of consciousness.

These issues are important for the study of ecstatic states, for one major interpretation of such states is that they are forms of consciousness. The limitation of the study of such states to later written accounts of the experience (especially those recounted by disciples who are conservative believers and who omit controversial biographical information) has the same problems as Freud's study of the "secondary process" in dreams. For Freud, the dreams we remember are not the actual dream material—those are hidden by the mind's censorship. The function of the secondary process is to hide dream material from the ego, allowing the censored version to be remembered. This is a major problem of the constructivist position—it is dealing with the secondary process, the later translations and interpretations of events (often written by disciples and followers), and not the primary one.

Wallace also notes the problem of the study of ecstatic states in the field of Religious Studies, "the only academic departments that are prohibited by law from promoting the truth of their subject matter" as well as being "the only academic field in which it is commonly assumed that those who neither believe in nor practice their subject matter are better able to understand it and teach it than those who do."[22] He sees this as very limiting and argues in favor of broader perspectives and methodologies, which can include the analysis of introspective and contemplative experience.

A mingling of religion and science would be useful, especially from the earlier approach of science as exploration. In the introduction of this book, we discussed the scientists who were interested in issues of religious consciousness. While the mainstream of psychology does not focus on this exploration, there are groups for whom this question is of great interest. For humanistic and transpersonal psychology, ecstasy is a valuable part of human experience and a source of insight into the deeper nature of the self. Both of these forms of psychology have been marginalized in recent decades, much as the study of religious consciousness has been marginalized in the field of religion. They both deserve another look, as useful methodologies for Religious Studies.

But including such views can be a challenge. There is a certain academic form of numinous dread that some writers in the field seem to feel toward scholars who value mystical and ecstatic states. One of the most articulate and extended arguments against mystical ecstasy as a major aspect of religion is found in Steven Wasserstrom's book *Religion After Religion*, in which he discusses the views of Gershom Scholem, Mircea Eliade, and Henry Corbin. As this book has been widely quoted as an argument against valuing religious ecstasy and mystical states, I will discuss it in some detail. Wasserstrom goes over the views of Scholem, Eliade and Corbin, and he describes his objections to their view of the importance of mystical experience in the understanding of religions. He states,

> In this theoretical revision, each scholar underplayed the importance of law, ritual and social history. Instead they primarily were concerned with myth and mysticism. With this striking reversal of the conventional emphases of adherents as well as of many scholars, they developed a monotheism without ethics.[23]

He describes this as the problem of the opposition between classical and romantic views of the world, and scholars must follow the classical perspective, for ethics are more important than transcendence. If scholars focus on mysticism and religious symbolism, then he assumes that all that they can achieve is an exploration of the aesthetic, which he finds to be less valuable. It takes away from the focus on the physical world, which is the more important locale. As Wasserstrom notes, "Such presumptions of an archetypal reality more real than ordinary reality tended to bleed the ordinary of its interest and vigor."[24] Clearly, mystical goals and earthly ones are in a competition, and the scholar must support his team.

A focus on mystical and ecstatic states can question the value of secular scholarship, a questioning which Wasserstrom does not like. As he writes of Eliade, "The romantic passion with which he wrote on questions of transcendence seemed to condescend, almost grudgingly, to the ordinary tolerations of scholarly life."[25] Here, Wasserstrom's argument is that the ordinary practitioners and scholars are the true elites, not the mystics who claim special status. Mystics and ecstatics are just making false claims. Indeed, as he further states (italics in the original), "*the esoteric art of writing is, in plain language, also a form of lying.*"[26] For him, the use of esoteric symbols is "camouflage, or heavenly deception."

If something cannot be expressed in ordinary language, its "dangerous appeal" is dishonest. The use of mystical symbolism is only aesthetic, rather than meaningful. The History of Religions must focus itself on the concerns of the many rather than the few; it therefore should not discuss esoteric knowledge and thus support the "rejection of the self-understanding of millions of ordinary believers."[27]

He was also concerned with the arguments in favor of wholeness by the members of the Eranos group, primarily Carl Jung, as well as others who followed him. Though he admitted that the early field of History of Religions was marked by "Nietzschean intensity and by monographs bristling with extreme formulations, grandiose projects and pyrotechnic displays of erudition,"[28] still they were lacking in morality, for they were interested in the possible integration of evil into religion. As chapter titles like "On the Suspension of the Ethical," "Moral Authority without Ethical Monotheism," "A Sort of Moral Conspiracy," and "The Apparent Absence of Ethics at Eranos" show, Wasserstrom objected to a "godhead of transmoral infinity" and emphasized a God who gave commandments and people following those commandments. Symbolism made commandments appear less important, and this corrupted religions. Seeking to experience God more directly becomes a threat to ethics.

Spiritual freedom without ethical restraint meant transgression, leading to a nihilistic emphasis on corruption and degeneracy. The ontical and the aesthetic could not legitimately replace the centrality of the ethical in History of Religions, and he suggests that Eliade, Scholem, and Corbin followed nihilistic mysticism as a sort of antiethic.[29] He states that ethics break down when the world is deconstructed in mystical experience, and thus, the mystic or gnostic is "an exception to the rules binding the rest of us." The individual encounter or theophany should not be more important than God's "unequivocally commanding ethical voice." As for the field of History of Religions, it connects us to the religious past. As Wasserstrom notes, "This past was not crowded with mystics, but rather with detailed practices and personalities of all types."[30] He seems to have missed many practitioners in this view.

He notes that much of the spirit and enthusiasm of the older forms of History of Religions has fled to the New Age, as well as to the arts. He does not consider this a problem—it did not belong in the academy anyway. He describes religious experience as marginalized, just as Continental Philosophy has become marginalized in philosophy departments. In his subchapter "On Mystocentrism," he condemns writers who find the essence of religion in mystical experience. Historians of religion should study only texts and contexts, not religious phenomena, and such metaphysical concerns should not be "urged on innocent students." Sometimes he describes the Historians of Religion as wanting people to only study mystical issues, sometimes as primarily doing so. In response, he wants the topic to various degrees out of the academy; thus, he emphasizes its "inadequacies." History of Religions should focus on the public life of believers.[31]

What we see here is a collection of value judgments and ethical norms which become authoritative statements about what a field should be. Indeed, Wasserstrom values the commandments, but where would the commandments be if Moses had not had his theophany? Even if the commandments suddenly appeared from a secular nowhere, why would people value them? Should the ethical norms derived from religious texts be the only allowable topic of religious study? This narrows the field of legitimate knowledge considerably. Without the authority that comes from revelatory religious experience, religious ethics have no firm foundations, to support the civic virtues and social stability he values.

A very different view comes from Philip Wexler. While he too is interested in the public life of believers, he does not understand mysticism to be an enemy of society. His perspective is a more pragmatic one, arguing that a value on mystical and ecstatic states can balance the "apparel limitless excesses" found in modern capitalist societies. These include greed, isolation, alienation, and objectification. He is in favor of a "remysticization of society," which can counter the "exhaustion and routinization of ecstasy" that we see today, and bring a renewed sense of depth and meaning to life.[32] Following Thomas Csordas, he supports a "transnational transcendence" which could revitalize modern culture and redefine the self by including a dimension of sacredness.[33]

For Wexler, the current "spiritual turn" in some areas of society would also include a revitalization of academic mystical scholarship.[34] As a sociologist, he is particularly interested in developing a field of mystical sociology which would be "a hybridization of sacred and secular." However, he is also interested in a broader focus in the humanities and social sciences, where the perspective of remysticization could become a common resource for scholars. As postmodern spirituality has become very diverse, becoming a matter of choice and creativity and self-expression as well as tradition, finding deeper unities becomes an important strategy.

There are some directions which hold promise for the study of religious ecstasy in the fields of History of Religions and Religious Studies. The methodology of Anthropology of Religion allows for the ethnography and cultural sensitivity that we often find in the study of Eastern religions, especially in the study of India. It is an older direction whose values have been lost in the modern arguments over ideology and power, yet many researchers have shown sensitivity to the religious worlds of their informants. The field of Religious Studies would be well served by more empathic ethnographic research.[35]

This emphasis upon the religious experiences and understandings of informants has been brought more recently to the study of Western religions, especially in the area of "lived religion." One researcher who holds this approach is Robert Orsi. He defines lived religion as including "the work of social agents/actors themselves as narrators and interpreters (and reinterpreters) of their own experiences and histories, recognizing that the stories we tell about others exist alongside the many and varied stories they tell of themselves."[36] Orsi describes lived religion as centered on the actions and interpretations of religious people. His emphasis on non-traditional forms of research encourages scholars to give attention to institutions and persons, texts and rituals, practices and Theology, things and ideas, human actors, and supernatural ones. In order to study lived religion, Orsi suggests a complex academic lens where a wide range of experiences can hold meaning and serve as a source or text for study. Thus, experiences with saints and angels can be as important as conversations with priests and theologians.

More specifically, Orsi defines religion as "a network of relationships between heaven and earth" rather than a medium for explaining and modeling reality.[37] What is important is intersubjectivity rather than ethicizing. The tendency to ethicize and create religious norms historically has meant excluding religious approaches that Western scholars did not appreciate, and it brought in racist discourse and condemnation of groups that did not fit the norm. As Orsi phrases it,

> Rather, the entire notion of "religion" had been carefully demarcated to preserve it from ambivalence and ambiguity, from anything not in accordance with certain sanctioned notions of self and society. Religion came to be gridded along a graph of diametric opposites and the possibility that religion can transgress these various dualities, that it does its cultural, psychological and political work precisely by disregarding boundaries between

one self and another, or between past present and future, or between the natural and the supernatural, is disallowed.[38]

Religion as an ethical system was good for modern democracy and good for "domesticated, modern, civic Protestantism." Outside of the academy, the winds of religious madness howled—the "fire-baptized people, ghost dancers, frenzied preachers and gullible masses"—but religion "as it took shape in the academy was explicitly imagined in relation to these others and as a prophylactic against them."[39]

The focus on religious ethics narrows a field that should be a broad one. As Orsi states, "Any approach to religion that foregrounds ethical issues as these are now embedded in the discipline obstructs our understanding of religious idioms because religion at its root has nothing to do with morality."[40] For Orsi, religion should not only mean "desirable religion." It needs to include all aspects of religion, and all varieties, taking into account the experiences of the believers as well as the values of the outside observer. Religion involves "the realness of sacred presence in the imagination and experiences of religious practitioners."[41] However, it is a problem of modernity that such events are not only excluded, but hated, for there is "a deep antipathy between modern cultures all over the world and the practice and experience of sacred presence." He notes that "the modern world has assiduously and systematically disciplined the senses not to experience sacred presence; the imaginations of moderns are trained towards sacred absence."[42] Encounters with supernatural and divine beings, and visions of heavenly worlds, have an "unsavory and disreputable aura," which makes their study difficult. But they must be understood if we wish to understand the lives and values of our informants. As he states, "The saints, gods, demons, ancestors, and so on are real in experience and practice, in relationships between heaven and earth, in the circumstances of people's lives and histories, and in the stories people tell about them."[43] It is an approach which respects both supernatural experiences and material lives, without limiting truth to political conflicts.

Avoiding religious experience, and focusing on ethical norms, is the field of Religious Studies' way of limiting complexity, excluding marginal groups, and imposing closure. The field needs to include the otherness of religious belief and practice, without imposing moral and political views that mask the common humanity involved in beliefs and experiences outside the norm. Orsi writes that currently personal experience must be excluded from the modern field of Religious Studies, and this does not add to our knowledge: "It demands the intellectual and religious deformation of scholars who believe as the condition for admission to the guild."[44] He suggests a "third way," in which both observer and participant can have a role in academic knowledge. This approach would be transformative, able to create a kind of empathic communication rarely seen today, and negotiate different universes of meaning. Scholars should leave behind the security and moral certainty of their disciplines, in order to find new ways of understanding religious complexity.

Such a "third way" would be a useful addition to our toolbox of Religious Studies methodologies. But for now, the modern field of Religious Studies tends to view religious ecstasy with ambiguity and suspicion at best, and hostility at worst. In this way, it echoes the values of both mainstream analytic philosophy and mainstream cognitive psychology. Such a view separates ecstatic states from acceptable religious behavior, forcing it to enter other fields. Thus, we have ecstatic joy in destruction and violence, extreme sports, ecstatic music and dance, fascination with food, gambling and sex, and the religious ecstasy that may come at the end of the life of a religious martyr or someday at the end of the world if we follow commandments and God's will, but no ecstasy in modern life. If people want religious ecstasy now, they must leave the study and practice of mainstream religion for the vagueness of the New Age or enter the category of Spiritual but not Religious. While academics do not tend to approve the lack of serious study in these areas, they are forcing their own students in these directions, if the students wish to understand where their own life-changing experiences belong.

Not all people seek out ecstasy. There are many people happy to live out their lives without great passions and extreme experiences. But for those who value ecstasy, it is important to have guidance and ways to interpret such experiences. Without models and understandings of the dynamics involved, ecstatic states can be erratic, ending up in extremes of greed, destruction, and competition. With models and guidance, we can have intense experiences channeled toward ends that are more helpful to both the individual and the culture, emphasizing the ecstasy of religious and creative insight rather than the ecstasy of violence.

We return to the quote from David Steindl-Rast that began this book, "Every religion seems to begin in ecstasy and end in politics." From this perspective, we seem to be in the end-times. But perhaps a return to an interest in ecstatic origins could be a useful thing, as part of a broader study of religious consciousness, which is slowly returning to the other fields that have expelled it. Legitimate topics of study come into and out of fashion in academic fields. It is time for religious ecstasy to return as a subject of serious study in the field of religion.

Notes

1. See Jacque Berlinerblau's article on "The Crisis in Secular Studies" from 2014 in *The Chronicle of Higher Education*, at https://www.chronicle.com/article/The-Crisis-in-Secular-Studies/148599.
2. Paul Ricoeur, *The Symbolism of Evil*, trans. Emerson Buchanan (Boston: Beacon, 1969), p. 349.
3. Ibid., p. 350.
4. Ibid., p. 352.
5. Victoria Nelson, *The Secret Life of Puppets* (Cambridge: Harvard University Press, 2001), p. viii.
6. Ibid.
7. Elizabeth Lesser, *The New American Spirituality: A Seeker's Guide* (New York: Random House, 1999), p. 38.
8. Carol de Chellis Hill, *Henry James' Midnight Song* (New York: W.W. Norton, 1993), p. 19.
9. Jeffrey J. Kripal, *Authors of the Impossible: The Paranormal and the Sacred* (Chicago: University of Chicago Press, 2010), p. 26.
10. Ibid., p. 254.
11. Ibid., p. 6.
12. Paul Heelas, *The New Age Movement: The Celebration of the Self and the Sacralization of Modernity* (Oxford: Blackwell, 1996), p. 89.
13. Ibid., p. 217.
14. See Paul Heelas and Linda Woodhead, *The Spiritual Revolution: Why Religion is Giving Way to Spirituality* (Malden: Blackwell, 2005) and also Christopher Partridge, *The Re-enchantment of the West: Alternative Spiritualities, Sacralization, Popular Culture and Occulture* (London: Bloomsbury Academic, 2005).
15. Marianna Torgovnick, *Primitive Passions: Men, Women and the Quest for Ecstasy* (New York: Alfred A. Knopf, 1997), p. 219.
16. B. Alan Wallace, *The Taboo of Subjectivity: Toward a New Science of Consciousness* (New York: Oxford University Press, 2000), p. 5.
17. Ibid., pp. 27–28.
18. Cited in Wallace, ibid., p. 28.
19. Ibid., p, 29.
20. Ibid., p. 75.

21. Ibid., pp. 78–81.
22. Ibid., p. 170.
23. Steven M. Wasserstrom, *Religion After Religion: Gershom Scholem, Mircea Eliade, and Henry Corbin at Eranos* (Princeton: Princeton University Press, 1999), p. 5.
24. Ibid., p. 30.
25. Ibid., p. 24.
26. Ibid., p. 154.
27. Ibid., p. 182.
28. Ibid., p. 216.
29. Ibid., p. 230.
30. Ibid., p. 238.
31. Ibid., p. 241.
32. Philip Wexler, *Mystical Sociology: Toward Cosmic Social Theory* (New York: Peter Lang, 2013), p. 31.
33. Thomas Csordas, *Transnational Transcendence: Essays on Religion and Globalization* (Berkeley, CA: University of California Press, 2009).
34. Wexler, ibid., pp. 34–35.
35. As fine examples of ethnographic sensitivity in Indian traditions, we might include Antoinette DeNapoli, *Real Sadhus Sing to God: Gender, Asceticism and Vernacular Religion in Rajasthan* (New York: Oxford University Press, 2014); Margaret H. Case, *Seeing Krishna: The Religious World of a Brahman Family in Vrindaban* (New York: Oxford University Press, 2000); and M. Whitney Kelting, *Singing to the Jinas: Jain Laywomen, Mandal Singing, and the Negotiations of Jain Devotion* (Albany: SUNY Press, 2007). We could also include some edited volumes: Tracy Pintchman, ed., *Women's Lives, Women's Rituals in the Hindu Tradition* (Oxford University Press, 2007) and Selva Raj and William Harman, eds., *Dealing with Deities: The Ritual Vow in South Asia* (SUNY Press, 2006).
36. Robert Orsi, *The Madonna of 115th Street: Faith and Community in Italian Harlem, 1880–1950* (Yale University Press, 2002), p. xxxix.
37. Robert R. Orsi, *Between Heaven and Earth: The Religious Worlds People Make and the Scholars Who Study Them* (Princeton: Princeton University Press, 2005), p. 2.
38. Ibid., p. 186.
39. Ibid.
40. Ibid., p. 191.
41. Ibid., p. 10.
42. Ibid., p. 12.
43. Ibid., p. 18.
44. Ibid., p. 193.

Index

J. McDaniel, *Lost Ecstasy*, Interdisciplinary Approaches to the Study of Mysticism, https://doi.org/10.1007/978-3-319-92771-8

The manufacturer's authorised representative in the EU is Springer Nature Customer Service Centre GmbH, Europaplatz 3, 69115 Heidelberg, Germany. If you have any concerns regarding our products, please contact ProductSafety@springernature.com

Printed and bound by CPI Group (UK) Ltd, Croydon, CR0 4YY
05/10/2026
02228128-0010